INNOVATIONS THAT CHANGED THE WORLD

A Journey through Science and Technology

SREEKUMAR V T

PREFACE

In the vast tapestry of human history, few threads are as vibrant and transformative as those woven by the ingenuity of science and technology. From the earliest sparks of curiosity that led our ancestors to create tools from stone, to the sophisticated advancements that drive today's digital age, the journey of innovation has been both a testament to human potential and a catalyst for profound change.

"Innovations That Changed the World: A Journey through Science and Technology" is an exploration of this remarkable journey. This book seeks to illuminate the pivotal discoveries and inventions that have not only shaped our past but continue to influence our present and future. Each chapter delves into a different era or field of innovation, highlighting the brilliant minds and groundbreaking achievements that have propelled humanity forward.

The story of science and technology is one of relentless curiosity, tireless experimentation, and boundless creativity. It is a story of overcoming challenges, pushing boundaries, and imagining possibilities beyond the horizon. It is also a story of interconnectedness, where each discovery builds upon the foundations laid by previous generations, creating a cumulative legacy of knowledge and progress.

As you turn these pages, you will encounter tales of astonishing breakthroughs and the individuals who dared to dream them. You will witness the profound impact of these innovations on society, culture, and everyday life. From the humble beginnings of ancient inventions to the awe-inspiring advancements of the modern era, this book offers a comprehensive overview of how science and technology have continually reshaped our world.

This journey through time and innovation is not just a celebration of human achievement; it is also a reminder of the potential that lies within each of us to contribute to the ever-evolving story of progress. Whether you are a student of science, a history enthusiast, or simply curious about the forces that have shaped our world, this book is an invitation to explore, to learn, and to be inspired.

I hope that as you read this book, you will gain a deeper appreciation for the marvels of science and technology and the incredible journey of human innovation. May it inspire you to look at the world around you with a sense of wonder and to imagine the endless possibilities that the future holds.

Welcome to "Innovations That Changed the World: A Journey through Science and Technology."

Sincerely,

SREEKUMAR V T

COPYRIGHT WARNING

Copyright © [2024] [Sreekumar V T]

All rights reserved. No part of this publication may be reproduced, distributed, or transmitted in any form or by any means, including photocopying, recording, or other electronic or mechanical methods, without the prior written permission of the publisher, except in the case of brief quotations embodied in critical reviews and certain other non-commercial uses permitted by copyright law.

The unauthorized reproduction or distribution of this copyrighted work is illegal. Criminal copyright infringement, including infringement without monetary gain, may be investigated and prosecuted by federal authorities.

For permission requests, please contact the publisher at the email address vtsreekumar123@gmail.com

CONTENTS

1. The Dawn of Discovery: Ancient Inventions and Their Legacy

2. Navigating the Unknown: The Age of Exploration and Maritime Technology

3. The Spark of Genius: The Scientific Revolution

4. Powering the Future: The Industrial Revolution and Mechanization

5. The Rise of Communication: Telegraphs, Telephones, and the Internet

6. Medical Marvels: Breakthroughs in Health and Medicine

7. From the Earth to the Stars: Space Exploration and Astronomy

8. Digital Revolution: The Birth and Evolution of Computing

9. Energy Innovations: From Steam Engines to Renewable Power

10. Materials of Tomorrow: Advancements in Chemistry and Materials Science

11. Artificial Intelligence: The New Frontier of Technology

12. Sustainable Futures: Innovations in Environmental Science and Technology

1. THE DAWN OF DISCOVERY

Ancient Inventions and Their Legacy

The story of human progress is, at its core, a story of discovery and invention. From the earliest days of our species, our ancestors have sought to understand the world around them, to solve the problems they encountered, and to improve their lives through innovation. This chapter explores the dawn of discovery, a time when the seeds of scientific and technological advancement were first sown. It examines the ancient inventions that laid the groundwork for future progress and the enduring legacy of these early innovations.

The Spark of Curiosity

Long before the written word, before the rise of great civilizations, humans were already shaping their environment with tools and techniques born of necessity and curiosity. The earliest evidence of human ingenuity can be traced back to the Stone Age, a period that began around 2.5 million years ago. During this time, early humans developed simple tools made of stone, bone, and wood, marking the first steps in a journey that would lead to the sophisticated technologies of the modern world.

The creation of stone tools, such as hand axes and flint knives, represents one of the first major technological milestones in human history. These tools not only improved the ability of early humans to hunt and gather food but also facilitated the processing of animal hides and plant materials. The techniques used to produce these tools, involving the precise chipping and shaping of stone, required a high degree of skill and knowledge. This early craftsmanship laid the foundation for future technological developments and demonstrated the capacity for complex thought and problem-solving.

The Agricultural Revolution

One of the most significant periods in human history is the Agricultural Revolution, which began around 10,000 BCE. This era marked the transition from a nomadic, hunter-gatherer lifestyle to settled agricultural communities. The development of agriculture was a pivotal innovation that transformed human societies in profound ways.

The domestication of plants and animals allowed for more reliable food sources, leading to population growth and the establishment of permanent settlements. With stable food supplies, communities could diversify their activities and develop specialized skills. This, in turn, led to the creation of new tools and technologies to support agricultural practices.

Irrigation systems, for example, were developed to manage water resources and ensure the successful cultivation of crops in arid regions. The ancient Mesopotamians, Egyptians, and Indus Valley civilizations engineered complex networks of canals, dikes, and reservoirs to control water flow and support their agricultural needs. These early irrigation systems required sophisticated planning and engineering, showcasing the advanced understanding of hydraulics and environmental management possessed by these ancient societies.

Writing and Record-Keeping

As agricultural societies grew more complex, the need for record-keeping and communication became increasingly important. The invention of writing, therefore, stands as one of the most transformative innovations of the ancient world. Writing systems allowed for the recording of information, the transmission of knowledge, and the administration of growing societies.

One of the earliest known writing systems is cuneiform, developed by the Sumerians of ancient Mesopotamia around 3400 BCE. Cuneiform script, inscribed on clay tablets, was used to document everything from agricultural inventories to legal codes. This ability to record and store information enabled the efficient management of resources and the codification of laws, contributing to the stability and growth of Sumerian civilization.

Similarly, ancient Egypt developed hieroglyphics, a complex system of pictorial writing, around 3200 BCE. Hieroglyphics were used for religious texts, monumental inscriptions, and administrative records. The preservation of knowledge through writing allowed for the continuity of cultural and scientific achievements across generations.

Mathematics and Astronomy

The development of mathematics and astronomy in ancient civilizations further illustrates the profound impact of early innovations on human progress. These fields of knowledge not only facilitated practical applications, such as construction and navigation, but also deepened the understanding of the natural world.

The ancient Egyptians, for instance, made significant contributions to mathematics, particularly in the fields of geometry and arithmetic. Their understanding of geometry was essential for the construction of monumental structures such as the pyramids. The precise alignment and proportions of these structures demonstrate an advanced knowledge of mathematical principles.

In Mesopotamia, the Babylonians developed a sophisticated numerical system based on the sexagesimal (base-60) system. This system facilitated complex calculations and was used in various applications, including astronomy. The Babylonians made remarkable astronomical observations, charting the movements of celestial bodies and developing early forms of calendars. Their work laid the groundwork for future astronomical discoveries and the development of more accurate timekeeping systems.

The Wheel and Transportation

The invention of the wheel is often hailed as one of the most significant technological advancements in human history. The wheel, believed to have been invented around 3500 BCE in Mesopotamia, revolutionized transportation and had far-reaching implications for trade, warfare, and daily life.

The earliest wheels were likely used for pottery making, but their application to transportation soon followed. The creation of wheeled vehicles, such as carts and chariots, greatly increased the efficiency of moving goods and people. This innovation facilitated trade over long distances, allowing for the exchange of goods, ideas, and cultural practices between different regions.

The impact of the wheel extended beyond transportation. It also influenced the development of other technologies, such as the potter's wheel, which enabled the mass production of pottery. The use of wheeled vehicles in warfare, particularly chariots, changed the dynamics of battle and contributed to the rise and fall of empires.

Metallurgy and Toolmaking

The mastery of metallurgy, the science of working with metals, marked another major milestone in ancient technological development. The discovery and utilization of metals such as copper, bronze, and iron had profound effects on toolmaking, construction, and warfare.

The transition from the Stone Age to the Bronze Age, around 3000 BCE, was characterized by the use of bronze, an alloy of copper and tin, to create stronger and more durable tools and weapons. Bronze tools improved agricultural practices, construction techniques, and craft production. The ability to produce metal weapons also altered the balance of power between communities and played a crucial role in the expansion of ancient civilizations.

The subsequent advent of the Iron Age, around 1200 BCE, further revolutionized toolmaking and warfare. Iron, being more abundant and harder than bronze, allowed for the creation of superior tools and weapons. The widespread use of iron technology contributed to significant societal changes, including the rise of powerful empires and the spread of advanced agricultural practices.

Engineering Marvels

The engineering achievements of ancient civilizations continue to inspire awe and admiration. The construction of monumental structures, such as the pyramids of Egypt, the ziggurats of Mesopotamia, and the aqueducts of Rome, showcases the remarkable ingenuity and technical prowess of ancient engineers.

The Great Pyramid of Giza, built around 2580–2560 BCE, remains one of the most iconic and enduring symbols of ancient engineering. The precise construction and alignment of this massive structure, composed of millions of limestone blocks, required meticulous planning, labor, and knowledge of geometry and astronomy. The pyramid's construction techniques continue to be studied and debated by scholars and engineers today.

In Mesopotamia, the construction of ziggurats, massive terraced structures, demonstrated advanced architectural and engineering skills. These structures, such as the Ziggurat of Ur, served religious and administrative purposes and required the coordination of large workforces and resources.

The Romans, renowned for their engineering prowess, developed a vast network of roads, aqueducts, and public buildings. Roman aqueducts, such as the Pont du Gard in France, exemplify the advanced understanding of hydraulics and construction techniques. These aqueducts transported water over long distances, supporting the growth and sustainability of Roman cities.

Medicine and Healing

The practice of medicine in ancient civilizations was deeply intertwined with early scientific and technological innovations. Ancient medical practices combined empirical observations with religious and philosophical beliefs, leading to the development of various treatments and surgical techniques.

In ancient Egypt, medical knowledge was highly advanced for its time. The Edwin Smith Papyrus, dating to around 1600 BCE, is one of the earliest known medical texts. It contains detailed descriptions of surgical procedures, treatments for injuries, and anatomical observations. Egyptian physicians practiced techniques such as suturing wounds, setting fractures, and performing complex surgeries.

The ancient Greeks made significant contributions to the field of medicine, particularly through the work of Hippocrates and his followers. Hippocrates, often regarded as the "father of medicine," emphasized the importance of observation and diagnosis. The Hippocratic Corpus, a collection of medical texts, laid the foundation for modern medical ethics and practices.

The ancient Chinese also developed a rich tradition of medical knowledge, exemplified by texts such as the Huangdi Neijing (The Yellow Emperor's Classic of Internal Medicine). This text, dating to around 300 BCE, covers various aspects of traditional Chinese medicine, including acupuncture, herbal remedies, and the concept of qi (vital energy). The holistic approach to health and wellness in ancient Chinese medicine continues to influence contemporary medical practices.

The Legacy of Ancient Innovations

The innovations of ancient civilizations have left an enduring legacy that continues to shape our world today. These early discoveries and inventions provided the foundation for subsequent advancements in science and technology, influencing countless aspects of modern life.

The principles of mathematics and engineering developed by ancient societies are still applied in contemporary construction, architecture, and technology. The agricultural practices and irrigation techniques pioneered by ancient civilizations continue to inform modern farming and water management strategies.

The invention of writing and record-keeping enabled the preservation and transmission of knowledge across generations, laying the groundwork for the development of complex societies and cultures. The medical practices and anatomical knowledge of ancient healers laid the foundation for modern medicine and surgery.

Moreover, the spirit of curiosity, experimentation, and innovation that drove ancient discoveries remains a driving force behind scientific and technological progress today. The achievements of ancient inventors serve as a reminder of the boundless potential of human ingenuity and the power of knowledge to transform the world.

As we embark on this journey through the history of science and technology, it is essential to recognize and appreciate the contributions of ancient civilizations. Their innovations, born of necessity and curiosity, have had a profound and lasting impact on the trajectory of human progress. By understanding and building upon their legacy, we continue to push the boundaries of what is possible, exploring new frontiers and shaping the future of our world.

2. NAVIGATING THE UNKNOWN

The Age of Exploration and Maritime Technology

The Age of Exploration, spanning from the 15th to the 17th centuries, was a transformative period in human history marked by significant advancements in maritime technology and the pursuit of new frontiers. This era saw the rise of intrepid explorers who navigated uncharted waters, established trade routes, and connected distant continents. The innovations in shipbuilding, navigation, and cartography during this time not only facilitated these voyages but also laid the groundwork for the modern globalized world. This chapter delves into the technological advancements that made these explorations possible and the profound impact they had on the course of history.

The Impetus for Exploration

Several factors drove the Age of Exploration, including economic interests, political ambitions, and a thirst for knowledge. The desire to find new trade routes to Asia, particularly to access valuable spices, silk, and other goods, was a primary motivator. European powers, especially Portugal and Spain, were eager to bypass the overland routes controlled by Ottoman and Arab traders, seeking direct access to these lucrative markets.

Religious zeal also played a significant role, as Christian nations aimed to spread their faith and counter the influence of Islam. Additionally, the

Renaissance, with its emphasis on discovery and learning, inspired a renewed interest in exploring the world and understanding its geography.

Innovations in Shipbuilding

The success of the Age of Exploration hinged on advancements in shipbuilding that produced vessels capable of long-distance sea travel. Two key innovations during this period were the development of the caravel and the galleon.

The Caravel

The caravel, a small, highly maneuverable ship developed by the Portuguese in the 15th century, was instrumental in early exploration efforts. Caravels were designed with a combination of square and lateen sails, which provided both speed and the ability to sail windward. This versatility allowed explorers to navigate the challenging winds and currents of the Atlantic Ocean.

Notable features of the caravel included its shallow draft, which enabled it to sail close to shore and explore rivers and estuaries, and its strong yet lightweight construction, which made it durable for long voyages. Famous explorers such as Bartolomeu Dias and Vasco da Gama utilized caravels in their groundbreaking journeys along the African coast and to India, respectively.

The Galleon

As exploration expanded, so did the need for larger, more robust ships capable of carrying substantial cargo and withstanding the rigors of the open sea. The galleon, developed in the late 16th century, met these needs. Galleons were multi-decked, square-rigged ships with a high forecastle and aftcastle, providing better stability and protection in rough waters.

Galleons were also heavily armed, with multiple gun decks that allowed them to serve both as trading vessels and warships. This dual purpose made them ideal for the increasingly competitive and dangerous waters of the Age of Exploration, where piracy and naval conflicts were common. The Spanish galleon, in particular, became synonymous with the treasure fleets that transported vast riches from the Americas to Europe.

Advances in Navigation

While improved ship designs were crucial, the ability to navigate vast, uncharted oceans was equally important. The Age of Exploration saw significant advancements in navigational instruments and techniques, many of which were built upon earlier innovations from the Islamic world and classical antiquity.

The Compass

The magnetic compass, which had been used in China since the 11th century, was adopted by European mariners in the 13th century and became an essential tool for navigation. By the time of the Age of Exploration, the compass had been refined and was widely used to determine direction at sea. The compass allowed sailors to maintain a consistent course, even when out of sight of land, reducing the risk of becoming lost in the open ocean.

The Astrolabe and Quadrant

The astrolabe, an ancient instrument used to measure the altitude of celestial bodies, was adapted for maritime use during the Age of Exploration. By determining the angle of stars above the horizon, sailors could calculate their latitude, which was crucial for open-ocean navigation. The quadrant, a simpler instrument used for similar purposes, also became a common tool on ships.

The Cross-Staff and Back-Staff

To further improve latitude measurement, navigators developed the cross-staff and later the back-staff. The cross-staff, consisting of a wooden rod with a sliding crosspiece, allowed sailors to measure the angle between the horizon and a celestial body. The back-staff, invented by John Davis in the late 16th century, improved upon this design by allowing measurements to be taken with the observer's back to the sun, reducing the risk of eye damage.

The Marine Chronometer

One of the greatest challenges in navigation was determining longitude, which required accurate timekeeping. The development of the marine chronometer in the 18th century, by John Harrison, marked a significant breakthrough. This highly precise clock could keep accurate time at sea, allowing sailors to calculate their longitudinal position by comparing the local time with the time at a known reference point, such as Greenwich.

Although the chronometer came slightly after the main period of the Age of Exploration, it was a culmination of efforts and technological advancements made during that era.

Cartography and Mapping

The advancements in navigation were complemented by significant progress in cartography. Accurate maps were essential for planning voyages, claiming territories, and establishing trade routes. The Age of Exploration saw the creation of detailed maps and charts that expanded the known world and facilitated further exploration.

Ptolemy's Influence

The works of the ancient Greek geographer Claudius Ptolemy, particularly his "Geographia," had a profound influence on Renaissance cartography. Rediscovered in the 15th century, Ptolemy's maps and geographical concepts provided a foundation for European mapmakers. Although his maps were not entirely accurate, they offered a systematic approach to mapping the world, which was built upon by explorers and cartographers of the Age of Exploration.

Portolan Charts

Portolan charts, developed in the 13th and 14th centuries, were among the earliest detailed maritime maps. These charts, characterized by their rhumb lines (lines of constant bearing), depicted coastlines, harbors, and navigational hazards with remarkable accuracy. Portolan charts were invaluable to early explorers, guiding them along established sea routes and helping them to avoid dangers.

The Age of Discovery Maps

The voyages of explorers like Christopher Columbus, Ferdinand Magellan, and James Cook contributed to the creation of new, more accurate maps. These explorers meticulously documented their journeys, providing valuable data that cartographers used to update and expand the known world. For instance, the Waldseemüller map of 1507 was the first to depict the New World as separate from Asia, reflecting the discoveries of Columbus and Amerigo Vespucci.

Mercator's Projection

One of the most significant contributions to cartography during the Age of Exploration was the Mercator projection, developed by Gerardus Mercator in 1569. This cylindrical map projection allowed for straight rhumb lines, which are essential for accurate navigation over long distances. The Mercator projection became the standard for nautical charts and remains widely used in maritime navigation today.

Key Explorers and Their Contributions

The Age of Exploration was marked by the exploits of daring explorers who ventured into unknown waters, driven by the promise of wealth, glory, and knowledge. Their voyages not only expanded the geographical horizons of their time but also facilitated cultural exchanges and the spread of technologies.

Christopher Columbus

Christopher Columbus, an Italian navigator sponsored by Spain, is perhaps the most famous explorer of this era. In 1492, Columbus embarked on his first voyage across the Atlantic, seeking a westward route to Asia. Instead, he stumbled upon the Americas, a "New World" that would profoundly alter the course of history. Although he never realized he had discovered a new continent, Columbus's voyages opened the way for European colonization and exploration of the Americas.

Vasco da Gama

Portuguese explorer Vasco da Gama made one of the most significant voyages of the Age of Exploration by reaching India by sea in 1498. His successful navigation around the Cape of Good Hope and across the Indian Ocean established a direct maritime route to Asia, bypassing the overland routes dominated by Muslim traders. This achievement not only secured a lucrative spice trade for Portugal but also marked the beginning of European dominance in the Indian Ocean.

Ferdinand Magellan

Ferdinand Magellan, a Portuguese explorer in the service of Spain, led the first expedition to circumnavigate the globe. Setting sail in 1519, Magellan's fleet navigated the treacherous straits at the southern tip of South America, now known as the Strait of Magellan, and crossed the vast Pacific Ocean.

Although Magellan himself was killed in the Philippines, his expedition completed the journey, demonstrating the feasibility of global circumnavigation and highlighting the vastness of the Earth.

James Cook

James Cook, an English navigator and cartographer, made significant contributions to the exploration of the Pacific Ocean in the 18th century. Cook's voyages to Australia, New Zealand, and the Hawaiian Islands provided detailed maps and extensive knowledge of these previously uncharted regions. His scientific approach to exploration, including the use of advanced navigational tools like the marine chronometer, set new standards for accuracy and reliability in mapping.

The Impact of Maritime Technology on Globalization

The advancements in maritime technology during the Age of Exploration had far-reaching consequences that reshaped the world in profound ways. The ability to undertake long-distance sea voyages facilitated the establishment of global trade networks, the exchange of ideas and cultures, and the expansion of European empires.

Trade and Commerce

The establishment of new trade routes revolutionized global commerce. European powers, particularly Spain and Portugal, profited immensely from the trade of spices, precious metals, and other goods from the Americas, Africa, and Asia. The influx of wealth from these colonies fueled economic growth and the development of banking and financial systems in Europe.

The Dutch East India Company and the British East India Company, established in the 17th century, exemplify the commercial impact of the Age of Exploration. These powerful trading companies dominated global trade, establishing monopolies and influencing political affairs in their regions of operation. Their fleets, equipped with the latest maritime technologies, controlled key trade routes and facilitated the flow of goods and resources across the world.

Cultural Exchange

The Age of Exploration also fostered significant cultural exchanges, as explorers encountered diverse civilizations and brought back new ideas, technologies, and customs. The Columbian Exchange, named after Christopher Columbus, refers to the widespread transfer of plants, animals, people, and diseases between the Americas, Europe, Africa, and Asia.

This exchange had profound effects on societies across the globe. European diets were transformed by the introduction of crops like potatoes, tomatoes, and maize from the Americas. Conversely, Old World crops such as wheat, rice, and sugarcane were introduced to the New World. The spread of diseases like smallpox, however, had devastating impacts on indigenous populations in the Americas, leading to significant demographic and social changes.

Colonialism and Empire

The Age of Exploration paved the way for European colonialism, as powerful nations sought to expand their territories and exploit the resources of newly discovered lands. The establishment of colonies in the Americas, Africa, and Asia brought wealth and power to European empires but also led to the subjugation and exploitation of indigenous peoples.

The legacy of colonialism is complex and contentious. While it facilitated the spread of technologies, ideas, and innovations, it also caused immense suffering and displacement for many native populations. The cultural and economic impacts of colonialism continue to shape global relationships and power dynamics to this day.

Scientific Advancements

The scientific knowledge gained during the Age of Exploration contributed to the advancement of various fields, including geography, biology, and astronomy. The meticulous observations and documentation made by explorers expanded the understanding of the natural world and provided valuable data for future scientific endeavors.

For example, the detailed maps and charts created during this period laid the foundation for modern cartography and geographical sciences. The collection of botanical and zoological specimens by explorers like James Cook and Charles Darwin enriched the study of biology and contributed to the development of evolutionary theory.

The Enduring Legacy of Maritime Innovation

The technological advancements of the Age of Exploration have left an enduring legacy that continues to influence modern maritime practices and global interactions. The innovations in shipbuilding, navigation, and cartography from this era set the stage for the development of modern shipping and navigation technologies.

The principles of ship design and construction developed during the Age of Exploration, such as the use of multiple masts and advanced hull designs, are still applied in contemporary naval architecture. Modern ships, whether commercial cargo vessels or naval warships, owe much to the engineering achievements of this period.

Navigation technology has also continued to evolve, building on the foundations laid by early explorers. The development of GPS (Global Positioning System) and electronic navigation instruments has revolutionized maritime navigation, making it more precise and reliable than ever before. However, the basic principles of celestial navigation and the use of nautical charts remain essential skills for mariners.

The spirit of exploration and discovery that characterized the Age of Exploration continues to inspire new generations of explorers and scientists. The desire to push the boundaries of knowledge and explore uncharted territories is reflected in modern endeavors such as space exploration and deep-sea research.

The Age of Exploration was a pivotal period in human history, marked by remarkable advancements in maritime technology and a fervent quest to navigate the unknown. The innovations in shipbuilding, navigation, and cartography enabled explorers to venture into uncharted waters, establish new trade routes, and connect distant continents. These achievements not only reshaped the world of their time but also laid the groundwork for the modern globalized world.

The legacy of the Age of Exploration is multifaceted, encompassing the expansion of trade, cultural exchanges, colonialism, and scientific advancements. While it brought immense wealth and knowledge to European powers, it also had profound and often devastating impacts on indigenous populations and their cultures.

As we continue to explore new frontiers, whether in the depths of the oceans or the vastness of space, we are reminded of the enduring legacy of those who navigated the unknown during the Age of Exploration. Their spirit of curiosity, determination, and innovation continues to inspire and drive humanity's quest for discovery and understanding.

3. THE SPARK OF GENIUS

The Scientific Revolution

The Scientific Revolution, spanning roughly from the late 16th to the early 18th centuries, represents a profound transformation in human understanding and the way we perceive the natural world. This period was characterized by groundbreaking discoveries, the development of new methodologies, and a shift away from medieval scholasticism toward a more empirical and experimental approach to knowledge. The innovations and ideas that emerged during the Scientific Revolution laid the foundation for modern science and technology, shaping the intellectual landscape of the subsequent centuries. This chapter explores the key figures, discoveries, and the enduring legacy of this pivotal era.

The Intellectual Climate Before the Revolution

Before delving into the Scientific Revolution, it is important to understand the intellectual climate that preceded it. During the Middle Ages, European thought was dominated by scholasticism, a method of learning that emphasized the authority of classical texts and the teachings of the Church. Knowledge was largely derived from ancient authorities such as Aristotle and Ptolemy, and there was little emphasis on experimentation or empirical observation.

Medieval universities, while centers of learning, primarily focused on the study of theology, philosophy, and classical texts. The natural world was often understood through the lens of religious doctrine, and questioning established beliefs was discouraged. However, the Renaissance, which began in the 14th century, set the stage for the Scientific Revolution by reviving interest in classical knowledge and fostering a spirit of inquiry and humanism.

The invention of the printing press by Johannes Gutenberg in the mid-15th century played a crucial role in disseminating knowledge and ideas. Books became more accessible, enabling scholars to share their discoveries and build upon the work of others. This increased availability of information helped create an environment conducive to scientific advancement.

The Rise of Empiricism and the Scientific Method

One of the most significant developments of the Scientific Revolution was the emergence of the scientific method, a systematic approach to investigating natural phenomena. This method emphasized observation, experimentation, and the formulation of hypotheses. It marked a departure from reliance on authority and tradition, paving the way for a more rigorous and empirical approach to science.

Francis Bacon and the Inductive Method

Francis Bacon, an English philosopher and statesman, was a key figure in the development of the scientific method. In his work "Novum Organum," published in 1620, Bacon advocated for the use of inductive reasoning, which involves deriving general principles from specific observations. He argued that knowledge should be built upon empirical evidence rather than abstract reasoning or reliance on ancient authorities.

Bacon's emphasis on systematic observation and experimentation laid the groundwork for modern scientific inquiry. His ideas encouraged scientists to gather data through careful observation and to test their hypotheses through controlled experiments. This empirical approach became a cornerstone of the Scientific Revolution and continues to underpin scientific research today.

René Descartes and Deductive Reasoning

While Bacon championed inductive reasoning, René Descartes, a French philosopher and mathematician, emphasized the importance of deductive reasoning. In his work "Discourse on the Method," published in 1637, Descartes outlined a method of doubt and systematic skepticism. He argued that knowledge should be built upon clear and distinct ideas that could be deduced through logical reasoning.

Descartes' famous dictum, "Cogito, ergo sum" ("I think, therefore I am"), highlighted his belief in the certainty of self-awareness as the foundation of knowledge. His emphasis on rationalism and the use of mathematics to understand the natural world contributed to the development of the scientific method. Descartes' dualistic view of reality, separating mind and matter, also influenced subsequent scientific and philosophical thought.

Key Figures and Discoveries of the Scientific Revolution

The Scientific Revolution was marked by the contributions of numerous brilliant minds who made groundbreaking discoveries in various fields. These scientists challenged established beliefs, conducted innovative experiments, and developed new theories that transformed our understanding of the universe.

Nicolaus Copernicus and the Heliocentric Model

Nicolaus Copernicus, a Polish astronomer, is often credited with initiating the Scientific Revolution with his heliocentric model of the solar system. In his seminal work "De revolutionibus orbium coelestium" ("On the Revolutions of the Celestial Spheres"), published in 1543, Copernicus proposed that the Earth and other planets orbit the Sun, challenging the geocentric model of Ptolemy that had dominated Western thought for over a millennium.

Copernicus' heliocentric theory was revolutionary because it contradicted the widely accepted belief that the Earth was the center of the universe. Although his ideas were initially met with resistance, they laid the foundation for further astronomical discoveries and set the stage for a new understanding of the cosmos.

Galileo Galilei and the Telescope

Galileo Galilei, an Italian astronomer, physicist, and mathematician, made significant contributions to the Scientific Revolution through his use of the telescope. In 1609, Galileo constructed a telescope based on designs he had heard about and began observing the night sky. His observations provided compelling evidence for the heliocentric model and challenged prevailing notions about the nature of the universe.

Galileo discovered the moons of Jupiter, the phases of Venus, the rugged surface of the Moon, and the existence of countless stars in the Milky Way. His work "Sidereus Nuncius" ("Starry Messenger"), published in 1610, detailed these findings and argued against the Aristotelian view of a perfect, unchanging celestial realm.

Galileo's support for the heliocentric model brought him into conflict with the Catholic Church, leading to his trial and house arrest. Despite this, his contributions to astronomy, mechanics, and the scientific method were profound and enduring.

Johannes Kepler and the Laws of Planetary Motion

Johannes Kepler, a German mathematician and astronomer, built upon the work of Copernicus and Galileo to develop his laws of planetary motion. Kepler's meticulous analysis of astronomical data, particularly the observations of Mars by Tycho Brahe, led him to formulate three fundamental laws that described the motion of planets around the Sun.

Kepler's laws, published in "Astronomia Nova" (1609) and "Harmonices Mundi" (1619), demonstrated that planetary orbits are elliptical rather than circular, that planets sweep out equal areas in equal times, and that there is a precise mathematical relationship between the orbital period of a planet and its distance from the Sun. These laws provided a more accurate and predictive model of the solar system, further validating the heliocentric theory.

Isaac Newton and the Laws of Motion and Universal Gravitation

Isaac Newton, an English mathematician, physicist, and astronomer, is perhaps the most celebrated figure of the Scientific Revolution. His work "Philosophiæ Naturalis Principia Mathematica" ("Mathematical Principles of Natural Philosophy"), published in 1687, laid the foundation for classical mechanics and revolutionized our understanding of the natural world.

In the "Principia," Newton formulated the three laws of motion, which describe the relationship between the motion of an object and the forces acting upon it. These laws provided a comprehensive framework for understanding the behavior of objects in motion and formed the basis for Newtonian physics.

Newton's law of universal gravitation, also presented in the "Principia," posited that every mass in the universe exerts an attractive force on every other mass. This force, known as gravity, explained the motions of celestial bodies and unified the physics of the heavens and the Earth under a single theoretical framework.

Newton's contributions to mathematics, particularly the development of calculus, further underscored his profound impact on science. His work epitomized the empirical and mathematical rigor of the Scientific Revolution and set the stage for the scientific advancements of the Enlightenment and beyond.

The Impact on Other Scientific Disciplines

While astronomy and physics were central to the Scientific Revolution, other scientific disciplines also experienced significant advancements during this period. The application of the scientific method and the spirit of inquiry extended to fields such as biology, chemistry, and medicine.

Biology and Anatomy

The study of biology and anatomy underwent a transformation during the Scientific Revolution, driven by detailed observations and dissections. Andreas Vesalius, a Flemish anatomist, made significant contributions to the understanding of human anatomy through his meticulous dissections and observations. His seminal work "De humani corporis fabrica" ("On the Fabric of the Human Body"), published in 1543, challenged many of the inaccuracies found in the works of Galen, the ancient authority on anatomy.

Vesalius' detailed anatomical drawings and descriptions provided a more accurate representation of the human body, laying the groundwork for modern anatomy. His emphasis on direct observation and dissection as methods for studying the body exemplified the empirical approach of the Scientific Revolution.

Chemistry and Alchemy

The field of chemistry also evolved during the Scientific Revolution, moving away from the mystical and speculative practices of alchemy toward a more systematic and experimental approach. Robert Boyle, an Irish natural philosopher, is often regarded as one of the founders of modern chemistry.

His work "The Sceptical Chymist," published in 1661, criticized the traditional alchemical theories of matter and proposed a more rigorous and empirical method for studying chemical substances and reactions.

Boyle's experiments with gases led to the formulation of Boyle's Law, which describes the inverse relationship between the pressure and volume of a gas at constant temperature. His emphasis on experimentation, precise measurement, and the rejection of unsupported speculation were hallmarks of the new scientific approach.

Medicine and Physiology

The Scientific Revolution also brought significant advancements in medicine and physiology. William Harvey, an English physician, made groundbreaking discoveries in the circulatory system. His work "De Motu Cordis" ("On the Motion of the Heart and Blood"), published in 1628, demonstrated that blood circulates continuously through the body, pumped by the heart. This discovery challenged the prevailing Galenic model and provided a more accurate understanding of the cardiovascular system.

Harvey's meticulous observations and experiments exemplified the empirical methods of the Scientific Revolution. His work had a profound impact on the study of physiology and laid the foundation for modern medical science.

The Role of Institutions and Societies

The Scientific Revolution was not only driven by individual genius but also by the establishment of institutions and societies that fostered scientific inquiry and collaboration. These organizations provided platforms for the exchange of ideas, the dissemination of knowledge, and the validation of scientific discoveries.

The Royal Society

The Royal Society of London for Improving Natural Knowledge, founded in 1660, played a pivotal role in the advancement of science during the Scientific Revolution. The Royal Society promoted the systematic investigation of nature through observation and experimentation, embodying the principles of the scientific method.

Members of the Royal Society, including prominent figures such as Robert Hooke, Robert Boyle, and Isaac Newton, conducted experiments, published their findings, and engaged in lively discussions and debates. The society's motto, "Nullius in verba" ("Take nobody's word for it"), encapsulated its commitment to empirical evidence and independent verification.

The Académie des Sciences

The Académie des Sciences, established in Paris in 1666, was another influential scientific institution of the Scientific Revolution. Founded by Louis XIV and advised by the mathematician and philosopher René Descartes, the academy brought together leading scientists to collaborate on research and share their discoveries.

The Académie des Sciences supported a wide range of scientific disciplines, including astronomy, mathematics, physics, and natural history. Its members, such as Christiaan Huygens and Pierre-Simon Laplace, made significant contributions to their respective fields and helped advance the collective knowledge of the scientific community.

Universities and Academies

Universities and academies across Europe also played a crucial role in fostering scientific inquiry and education during the Scientific Revolution. Institutions such as the University of Padua, the University of Leiden, and the University of Cambridge became centers of scientific research and learning.

These universities attracted scholars from across Europe, facilitating the exchange of ideas and the development of new theories. The establishment of dedicated scientific chairs and laboratories within universities further supported the advancement of scientific knowledge.

The Legacy of the Scientific Revolution

The Scientific Revolution had a profound and lasting impact on the course of human history. Its legacy can be seen in the development of modern science, the advancement of technology, and the transformation of philosophical and intellectual thought.

The Rise of Modern Science

The Scientific Revolution laid the foundation for the rise of modern science by establishing the principles of empirical observation, experimentation, and mathematical reasoning. The scientific method, developed and refined during this period, became the standard approach for investigating the natural world and remains central to scientific inquiry today.

The discoveries and theories of the Scientific Revolution, such as the heliocentric model, the laws of motion, and the principles of gravitation, provided a coherent and predictive framework for understanding the universe. These advancements enabled subsequent generations of scientists to build upon this knowledge and make further breakthroughs in fields such as physics, chemistry, biology, and astronomy.

Technological Advancements

The innovations of the Scientific Revolution also had a direct impact on technological advancements. The development of accurate clocks, improved navigational instruments, and more precise measurement tools facilitated exploration, trade, and communication. These technologies contributed to the expansion of European empires, the growth of global commerce, and the spread of scientific knowledge.

The principles of mechanics and engineering developed during the Scientific Revolution also laid the groundwork for the Industrial Revolution, which began in the late 18th century. The application of scientific knowledge to practical problems led to the development of new machinery, manufacturing processes, and transportation systems that transformed economies and societies.

The Transformation of Intellectual Thought

The Scientific Revolution fundamentally transformed intellectual thought by challenging established beliefs and promoting a more critical and questioning approach to knowledge. The emphasis on empirical evidence and rational inquiry undermined the authority of traditional sources of knowledge, such as ancient texts and religious doctrine.

The ideas of the Scientific Revolution also influenced the Enlightenment, a cultural and intellectual movement of the 18th century that emphasized reason, individualism, and the pursuit of knowledge. Enlightenment thinkers, such as John Locke, Voltaire, and Immanuel Kant, drew upon the principles of the Scientific Revolution to advocate for political, social, and educational reforms.

The Pursuit of Knowledge

The spirit of inquiry and curiosity that characterized the Scientific Revolution continues to inspire scientists, researchers, and thinkers today. The pursuit of knowledge, driven by a desire to understand the natural world and improve the human condition, remains a fundamental aspect of the scientific endeavor.

The legacy of the Scientific Revolution is evident in the continued advancement of science and technology, from the exploration of space to the development of new medical treatments and the understanding of complex biological systems. The principles of empirical observation, experimentation, and rational inquiry established during this period continue to guide scientific research and drive innovation.

The Scientific Revolution was a transformative period in human history, marked by groundbreaking discoveries, the development of the scientific method, and the contributions of brilliant minds. This era laid the foundation for modern science and technology, shaping our understanding of the natural world and influencing subsequent intellectual and cultural developments.

The legacy of the Scientific Revolution is profound and enduring, as it established the principles of empirical observation, experimentation, and rational inquiry that continue to underpin scientific inquiry today. The advancements in astronomy, physics, biology, chemistry, and medicine during this period not only expanded our knowledge of the universe but also set the stage for future scientific and technological progress.

As we continue our journey through the history of science and technology, it is essential to recognize and appreciate the contributions of the Scientific Revolution. The spark of genius that ignited this era of discovery and innovation continues to inspire and drive humanity's quest for knowledge and understanding.

4. POWERING THE FUTURE

The Industrial Revolution and Mechanization

The Industrial Revolution, which began in the late 18th century and continued into the 19th century, was a period of profound transformation that reshaped economies, societies, and the very fabric of human life. It marked the transition from agrarian and handcraft-based economies to industrial and machine-based production. This era of innovation and mechanization brought about significant technological advancements, revolutionizing various industries and setting the stage for modern economic growth and development. This chapter explores the key innovations, figures, and the lasting impact of the Industrial Revolution.

The Beginnings of the Industrial Revolution

The Industrial Revolution began in Britain in the late 1700s and spread to other parts of the world over the following decades. Several factors contributed to its emergence in Britain, including the availability of natural resources, a stable political environment, a growing population, and an expanding colonial empire that provided raw materials and markets for manufactured goods.

Natural Resources and Geography

Britain's abundant natural resources, particularly coal and iron ore, played a crucial role in fueling the Industrial Revolution. Coal provided a reliable and efficient source of energy for powering steam engines and machinery, while iron ore was essential for constructing machines, tools, and infrastructure. The country's geographic features, including navigable rivers and a long coastline with natural harbors, facilitated the transportation of raw materials and finished goods.

Political and Economic Stability

Britain's political stability and economic policies also contributed to the onset of the Industrial Revolution. The country had a well-established banking and financial system, which provided capital for industrial investments. Additionally, the British government implemented policies that promoted trade and protected private property rights, encouraging entrepreneurship and innovation.

Population Growth and Urbanization

The population of Britain grew rapidly during the 18th century, providing a labor force for the emerging industries. This population growth, combined with agricultural improvements such as crop rotation and selective breeding, led to increased food production and a surplus labor force. Many people migrated from rural areas to urban centers in search of work, contributing to the growth of cities and the rise of industrial towns.

Colonial Empire and Trade

Britain's extensive colonial empire provided access to a wide range of raw materials, such as cotton from India and the American South, as well as markets for manufactured goods. The colonies also served as sources of wealth and investment capital, further fueling industrial development.

Key Innovations and Technological Advancements

The Industrial Revolution was characterized by a series of technological innovations that transformed manufacturing processes and increased productivity. These advancements laid the foundation for modern industry and mechanization.

The Textile Industry

The textile industry was one of the first sectors to undergo significant transformation during the Industrial Revolution. Several key inventions revolutionized textile production, making it more efficient and less labor-intensive.

The Spinning Jenny

The Spinning Jenny, invented by James Hargreaves in 1764, was a multi-spindle spinning frame that allowed a single worker to spin multiple threads simultaneously. This invention increased the output of yarn, reducing the need for skilled labor and lowering production costs.

The Water Frame

Richard Arkwright's Water Frame, developed in 1769, further revolutionized textile production. The Water Frame used water power to drive a series of spinning frames, enabling the production of stronger and finer yarn at a much larger scale. Arkwright's invention led to the establishment of large-scale factories, known as mills, along rivers and streams.

The Power Loom

The Power Loom, invented by Edmund Cartwright in 1785, mechanized the weaving process. This machine significantly increased the speed and efficiency of cloth production, allowing for the mass production of textiles. The widespread adoption of power looms led to the growth of the factory system and the decline of traditional handloom weaving.

Steam Power

The development of steam power was a pivotal moment in the Industrial Revolution, providing a new and powerful source of energy for industrial machinery and transportation.

James Watt and the Steam Engine

James Watt, a Scottish engineer, made significant improvements to the design of the steam engine in the 1760s and 1770s. Watt's enhancements, including the introduction of a separate condenser, increased the efficiency and reliability of steam engines. His partnership with entrepreneur Matthew Boulton led to the widespread adoption of steam engines in various industries, from textiles to mining.

The Steam Locomotive

The application of steam power to transportation was a major breakthrough. In 1804, Richard Trevithick built the first steam-powered locomotive, capable of hauling heavy loads over iron rails. However, it was George

Stephenson's locomotive, the Rocket, which achieved commercial success. In 1829, the Rocket won the Rainhill Trials, demonstrating its superior speed and reliability. The success of the Rocket led to the rapid expansion of railways, revolutionizing transportation and facilitating the movement of goods and people.

Iron and Steel Production

Advancements in iron and steel production were crucial to the Industrial Revolution, providing the materials needed for machinery, infrastructure, and construction.

The Blast Furnace

The development of the blast furnace, which used coke (a derivative of coal) instead of charcoal, allowed for the mass production of iron. This innovation, introduced by Abraham Darby in the early 18th century, significantly reduced the cost of iron production and increased its availability.

The Bessemer Process

In the mid-19th century, Henry Bessemer developed the Bessemer process, a method for producing steel by blowing air through molten iron to remove impurities. This process allowed for the mass production of steel, which was stronger and more durable than iron. The availability of affordable steel had a profound impact on construction, enabling the development of railways, bridges, and skyscrapers.

Mechanization and the Factory System

The introduction of machinery and the factory system transformed manufacturing, increasing efficiency and productivity.

The Factory System

The factory system, characterized by the use of machinery and the centralization of production, replaced the traditional cottage industry. Factories brought together large numbers of workers and machines under one roof, enabling the mass production of goods. This system led to economies of scale, lower production costs, and increased output.

Interchangeable Parts

The concept of interchangeable parts, popularized by Eli Whitney in the early 19th century, revolutionized manufacturing. By producing standardized parts that could be easily assembled and replaced, manufacturers improved efficiency and reduced production costs. This innovation laid the foundation for modern mass production techniques and the assembly line.

The Impact of the Industrial Revolution

The Industrial Revolution had far-reaching effects on various aspects of society, economy, and culture. It transformed the way people lived and worked, reshaped social structures, and had a profound impact on global trade and industry.

Economic Growth and Industrialization

The Industrial Revolution led to unprecedented economic growth and industrialization. The increased productivity and efficiency of manufacturing processes resulted in the mass production of goods, lowering prices and making products more accessible to a broader population. This economic growth fueled further investments in technology and infrastructure, creating a positive feedback loop of innovation and development.

Urbanization and Social Change

The rise of factories and industrial centers led to rapid urbanization, as people migrated from rural areas to cities in search of work. This migration resulted in the growth of industrial towns and cities, such as Manchester, Birmingham, and Liverpool in Britain.

Urbanization brought significant social changes, including the rise of a new working class and the development of new social dynamics. Factory work often involved long hours, low wages, and poor working conditions, leading to the emergence of labor movements and calls for social reform. The Industrial Revolution also contributed to the rise of the middle class, as increased economic opportunities allowed more people to achieve financial stability and social mobility.

Technological Innovation and Progress

The technological advancements of the Industrial Revolution set the stage for continued innovation and progress. The principles of mechanization, mass production, and the use of steam power were applied to various industries, from agriculture to transportation.

Agriculture

The Industrial Revolution had a significant impact on agriculture, as new machinery and techniques increased productivity and efficiency. The development of the mechanical reaper, invented by Cyrus McCormick in 1831, revolutionized grain harvesting, reducing labor costs and increasing output. The introduction of fertilizers and crop rotation also improved agricultural yields, supporting population growth and urbanization.

Transportation

The expansion of railways and the development of steam-powered ships revolutionized transportation, facilitating the movement of goods and people over long distances. The construction of canals and the improvement of road networks further enhanced connectivity and trade. These advancements laid the foundation for the modern transportation infrastructure that supports global commerce today.

Communication

The Industrial Revolution also saw significant advancements in communication technology. The invention of the electric telegraph by Samuel Morse in 1837 revolutionized long-distance communication, allowing for the rapid transmission of information. The development of the telephone by Alexander Graham Bell in 1876 further transformed communication, making it possible for people to speak with one another over long distances.

Global Trade and Colonial Expansion

The Industrial Revolution had a profound impact on global trade and colonial expansion. The increased production of goods and the demand for raw materials drove European powers to expand their colonial empires, seeking new markets and resources.

Trade Networks

The expansion of railways and steamships facilitated global trade, enabling the efficient transportation of goods between continents. The opening of the Suez Canal in 1869 and the Panama Canal in 1914 further enhanced global trade routes, reducing travel time and costs. These developments contributed to the growth of a global economy, with goods, capital, and labor flowing across borders.

Colonialism

The demand for raw materials, such as cotton, rubber, and minerals, fueled European colonial expansion in Africa, Asia, and the Americas. Colonial powers established plantations, mines, and trading posts, exploiting local resources and labor to supply their industrial economies. This expansion had significant economic, social, and political implications for the colonized regions, leading to the extraction of wealth and the disruption of traditional societies.

Environmental Impact

The Industrial Revolution had a profound impact on the environment, as the rapid expansion of industry and urbanization led to increased pollution and resource depletion.

Air and Water Pollution

The widespread use of coal as an energy source resulted in significant air pollution, with factories and steam engines releasing large quantities of smoke and pollutants into the atmosphere. This pollution contributed to health problems and environmental degradation, particularly in industrial cities.

Water pollution also became a major issue, as industrial waste and sewage were often discharged into rivers and streams. This contamination affected water quality and harmed aquatic ecosystems, leading to long-term environmental consequences.

Deforestation and Resource Depletion

The demand for raw materials, such as timber, coal, and minerals, led to extensive deforestation and resource depletion. Forests were cleared for agriculture, construction, and fuel, resulting in habitat loss and reduced biodiversity. The extraction of minerals and fossil fuels also had significant environmental impacts, including soil erosion, water contamination, and landscape alteration.

Labor Movements and Social Reforms

The harsh working conditions and social inequalities brought about by the Industrial Revolution led to the emergence of labor movements and calls for social reform.

Labor Unions

Labor unions, organizations formed by workers to advocate for their rights and interests, emerged during the Industrial Revolution. These unions fought for better wages, shorter working hours, and improved working conditions. Notable labor movements, such as the Chartist movement in Britain and the Knights of Labor in the United States, played a crucial role in advancing workers' rights and promoting social justice.

Social Legislation

The Industrial Revolution also prompted governments to enact social legislation aimed at addressing the negative impacts of industrialization. Laws regulating child labor, working hours, and factory conditions were introduced to protect workers and improve their quality of life. The Factory Acts in Britain, starting in 1802, set limits on working hours and conditions for children and women, marking the beginning of labor regulation.

Education and Innovation

The Industrial Revolution underscored the importance of education and innovation in driving economic growth and development. The need for skilled workers and engineers led to the establishment of technical schools and universities, fostering a culture of learning and innovation.

Technical Education

Technical education played a crucial role in supporting industrialization by providing workers with the skills and knowledge needed for new technologies and processes. Institutions such as the École Polytechnique in France and the Massachusetts Institute of Technology (MIT) in the United States were established to train engineers and scientists, contributing to technological advancements and industrial progress.

Innovation and Patents

The Industrial Revolution was a period of intense innovation, with inventors and entrepreneurs developing new technologies and processes to improve efficiency and productivity. The establishment of patent systems, such as the British Patent Office in 1852, encouraged innovation by providing legal protection for inventors' intellectual property. This system incentivized inventors to develop and share new ideas, contributing to the rapid pace of technological advancement.

The Legacy of the Industrial Revolution

The Industrial Revolution left an enduring legacy that continues to shape the modern world. Its impact can be seen in the technological, economic, and social developments that define contemporary society.

Technological Advancements

The technological innovations of the Industrial Revolution laid the groundwork for the continued advancement of science and technology. The principles of mechanization, mass production, and the use of new energy sources remain central to modern industry. The development of new materials, such as steel and plastics, and the invention of new machinery and processes have built upon the foundations established during this period.

Economic Transformation

The Industrial Revolution transformed economies by shifting the focus from agriculture to industry and manufacturing. This shift led to the growth of urban centers, the rise of a consumer society, and the development of global trade networks. The principles of industrialization, including the use of technology to increase productivity and efficiency, continue to drive economic growth and development.

Social and Cultural Changes

The Industrial Revolution brought significant social and cultural changes, including the rise of the middle class, the development of new social dynamics, and the emergence of labor movements. These changes have had a lasting impact on society, shaping modern social structures and influencing contemporary debates on workers' rights, social justice, and economic inequality.

Environmental Awareness

The environmental impact of the Industrial Revolution has also left a lasting legacy, raising awareness of the need for sustainable development and environmental protection. The negative consequences of industrialization, such as pollution and resource depletion, have highlighted the importance of balancing economic growth with environmental stewardship. This awareness has led to the development of environmental regulations, conservation efforts, and the promotion of sustainable technologies and practices.

The Industrial Revolution was a transformative period that revolutionized the way people lived and worked, reshaped economies and societies, and set the stage for modern technological and economic development. The innovations and advancements of this era, from mechanization and steam power to mass production and the factory system, had a profound and lasting impact on the world.

The legacy of the Industrial Revolution can be seen in the continued advancement of science and technology, the transformation of economies and societies, and the increased awareness of the need for sustainable development. As we continue to navigate the challenges and opportunities of the modern world, the lessons and innovations of the Industrial Revolution remain a vital source of inspiration and guidance.

5. THE RISE OF COMMUNICATION

Telegraphs, Telephones, and the Internet

Communication is the lifeblood of human society. From the earliest forms of spoken language to the sophisticated digital networks of today, the ability to share information has been a driving force behind cultural, social, and technological development. This chapter explores the transformative impact of three major innovations in communication: the telegraph, the telephone, and the Internet. Each of these technologies revolutionized the way people connect and interact, breaking down barriers of time and distance and paving the way for the globalized world we live in today.

The Telegraph: The Dawn of Electrical Communication

The invention of the telegraph in the early 19th century marked a pivotal moment in the history of communication. For the first time, messages could be sent across long distances almost instantaneously, transforming the speed and efficiency of information exchange.

The Birth of the Telegraph

Before the telegraph, communication over long distances was limited to physical means such as messengers, signal fires, and semaphore systems. These methods were slow and often unreliable, constrained by the limitations of human and animal endurance.

The concept of the electrical telegraph began to take shape in the early 19th century, building on the discoveries of electromagnetism by scientists such as Hans Christian Ørsted and André-Marie Ampère. In 1837, two British inventors, William Fothergill Cooke and Charles Wheatstone, developed the first practical telegraph system. Their device used multiple wires and a system of needles to point to letters on a board, allowing for the transmission of messages.

Samuel Morse and the Morse Code

While Cooke and Wheatstone's telegraph was groundbreaking, it was the American inventor Samuel Morse who truly revolutionized telegraphic communication. In 1838, Morse, along with his partner Alfred Vail, developed a simpler and more efficient telegraph system that used a single wire and a series of electrical pulses to transmit messages.

Morse also invented a binary code, known as Morse code, which represented letters and numbers using combinations of short and long signals (dots and dashes). This code allowed for the rapid and accurate transmission of messages, even over long distances.

The Expansion of Telegraph Networks

The first successful demonstration of Morse's telegraph took place in 1844, with the transmission of the famous message "What hath God wrought" from Washington, D.C., to Baltimore. This event marked the beginning of the telegraph's rapid expansion across the United States and Europe.

Telegraph companies quickly established extensive networks of wires, connecting cities, towns, and even remote areas. By the 1850s, the telegraph had become an essential tool for business, government, and the press. It enabled the rapid dissemination of news, facilitated commerce by allowing for real-time communication between distant markets, and played a crucial role in coordinating military operations.

The Transatlantic Telegraph Cable

One of the most ambitious projects of the telegraphic era was the laying of the transatlantic telegraph cable. In the mid-19th century, the idea of connecting Europe and North America with a submarine cable was seen as a monumental challenge due to the technical difficulties involved.

Cyrus West Field, an American entrepreneur, led the effort to lay the first successful transatlantic cable. After several failed attempts, the project finally succeeded in 1866. The transatlantic cable allowed for almost instantaneous communication between the continents, significantly reducing the time required for international correspondence and marking a major milestone in global connectivity.

The Impact of the Telegraph

The telegraph had a profound impact on society, transforming communication, commerce, and governance. It facilitated the rapid exchange of information, which was crucial for the functioning of financial markets, the coordination of transportation networks, and the management of large organizations.

The telegraph also played a critical role in journalism, enabling newspapers to report news from distant locations much more quickly than before. This revolutionized the news industry, leading to the rise of wire services such as the Associated Press.

In addition, the telegraph had significant social implications. It helped to shrink the world, making it possible for people to maintain personal and professional connections across great distances. The ability to communicate rapidly over long distances also had profound effects on diplomacy and international relations, contributing to the increasing interconnectedness of the world.

The Telephone: Bringing Voices Together

While the telegraph transformed written communication, the invention of the telephone brought about a revolution in voice communication. The ability to speak directly with someone far away added a new dimension to human interaction, making communication more immediate, personal, and effective.

Alexander Graham Bell and the Invention of the Telephone

The invention of the telephone is primarily attributed to Alexander Graham Bell, a Scottish-born inventor, scientist, and teacher of the deaf. Bell's interest in sound and speech led him to experiment with ways to transmit vocal sounds electrically.

On March 10, 1876, Bell successfully transmitted the first clear voice message using his telephone device. The message, "Mr. Watson, come here, I want to see you," was sent to his assistant, Thomas Watson, who was in another room. This moment marked the birth of a new era in communication.

The Development and Spread of Telephone Networks

Following Bell's groundbreaking invention, the telephone quickly gained popularity. The first commercial telephone exchange opened in New Haven, Connecticut, in 1878, allowing multiple subscribers to connect through a central switchboard operated by human operators.

The Bell Telephone Company, founded by Bell and his investors, played a crucial role in the development and expansion of telephone networks. The company built an extensive infrastructure of telephone lines and exchanges, making it possible for people to communicate over increasingly longer distances.

By the early 20th century, the telephone had become an integral part of daily life in many parts of the world. Urban areas were densely connected, and long-distance lines were being laid to connect cities and even countries.

Innovations in Telephone Technology

The telephone continued to evolve, with numerous innovations improving its functionality and accessibility. One significant advancement was the introduction of automatic switching systems, which replaced human operators with electromechanical devices capable of routing calls automatically. The first such system, invented by Almon Strowger, was patented in 1891 and gradually adopted by telephone companies.

The invention of the rotary dial in the early 20th century allowed users to dial numbers directly, further streamlining the process of making calls. This technology paved the way for the development of modern telephony.

The Impact of the Telephone

The telephone had a transformative impact on society, revolutionizing personal and business communication. It made it possible for people to have real-time conversations across long distances, fostering closer relationships and enhancing social connectivity.

In business, the telephone enabled more efficient communication, improving coordination and decision-making processes. It facilitated the growth of industries such as banking, insurance, and retail, which relied on timely and accurate communication.

The telephone also had significant cultural and social implications. It brought people closer together, reducing the isolation of rural areas and allowing families and friends to maintain contact despite geographical separation. The convenience and immediacy of voice communication also influenced social behaviors and expectations, shaping the way people interacted with one another.

The Internet: The Digital Revolution

The development of the Internet in the late 20th century marked the beginning of a new era in communication. This global network of interconnected computers revolutionized the way people access, share, and create information, transforming every aspect of modern life.

The Origins of the Internet

The origins of the Internet can be traced back to the 1960s, when the United States Department of Defense initiated research on a communication network that could withstand potential disruptions caused by a nuclear attack. This project, known as ARPANET (Advanced Research Projects Agency Network), laid the foundation for the modern Internet.

ARPANET was designed to connect computers at various research institutions, allowing them to share information and resources. The first successful message was sent over ARPANET in 1969, marking the beginning of computer networking.

The Development of TCP/IP and the Expansion of the Internet

One of the key developments in the evolution of the Internet was the creation of the Transmission Control Protocol/Internet Protocol (TCP/IP). This set of communication protocols, developed by Vint Cerf and Robert Kahn in the 1970s, provided a standardized way for different networks to communicate with each other. TCP/IP became the foundational protocol for the Internet, enabling the seamless exchange of data between diverse computer systems.

Throughout the 1980s and 1990s, the Internet expanded rapidly. The introduction of personal computers, modems, and networking technologies made it possible for individuals and businesses to connect to the Internet. The development of the World Wide Web by Tim Berners-Lee in 1989 further accelerated the growth of the Internet by providing a user-friendly interface for accessing and sharing information.

The World Wide Web and the Information Age

The World Wide Web (WWW), often simply referred to as the Web, revolutionized the way people interact with the Internet. Berners-Lee's creation, which included the concepts of hypertext and hyperlinks, allowed users to navigate the Internet through web pages and links. The first web browser, Mosaic, was released in 1993, making the Web accessible to a broader audience.

The Web transformed the Internet into a vast repository of information, enabling users to access a wide range of content, from academic research to entertainment. It also facilitated the development of e-commerce, social media, and online communication platforms, fundamentally changing the way people conduct business, socialize, and consume media.

The Impact of the Internet

The Internet has had a profound and far-reaching impact on virtually every aspect of modern life. It has transformed communication, commerce, education, entertainment, and many other fields, creating a highly interconnected and information-rich global society.

Communication and Social Connectivity

The Internet has revolutionized communication by enabling instant, global connectivity. Email, instant messaging, and video conferencing have made it possible for people to communicate in real time, regardless of geographical location. Social media platforms, such as Facebook, Twitter, and Instagram, have created new ways for people to connect, share, and engage with one another.

The rise of online communication has also facilitated the formation of virtual communities, allowing people with shared interests to interact and collaborate. This has had significant implications for social activism, political movements, and the spread of information.

E-Commerce and the Digital Economy

The Internet has transformed commerce by enabling online transactions and creating a global marketplace. E-commerce platforms, such as Amazon, eBay, and Alibaba, have revolutionized the way people buy and sell goods and services, offering convenience, variety, and competitive pricing.

The digital economy has also given rise to new business models and industries, such as digital marketing, cloud computing, and the gig economy. These developments have reshaped the global economy, creating new opportunities and challenges for businesses and workers.

Education and Knowledge Sharing

The Internet has democratized access to information and education, making it possible for people to learn and acquire knowledge from anywhere in the world. Online educational platforms, such as Khan Academy, Coursera, and edX, offer a wide range of courses and resources, enabling lifelong learning and professional development.

The availability of digital libraries, academic journals, and research databases has also transformed the way knowledge is created and shared. Researchers and students can access a vast array of information, collaborate with peers, and contribute to the global exchange of ideas.

Entertainment and Media

The Internet has revolutionized the entertainment and media industries, providing new ways for people to consume and create content. Streaming services, such as Netflix, YouTube, and Spotify, have changed the way people access and enjoy movies, music, and other forms of entertainment.

The rise of digital media has also transformed journalism and publishing. Online news outlets, blogs, and social media platforms have created new avenues for reporting and sharing information, challenging traditional media models and increasing the diversity of voices and perspectives.

The Future of Communication

The rapid evolution of communication technology continues to shape the future, with emerging technologies promising to further transform the way people connect and interact.

5G and Beyond

The rollout of 5G networks is expected to significantly enhance the speed, capacity, and reliability of mobile communication. This technology will enable the widespread adoption of the Internet of Things (IoT), connecting a vast array of devices and systems, from smart homes to autonomous vehicles. The increased connectivity and data capacity of 5G will also support the development of new applications and services, driving innovation across various industries.

Artificial Intelligence and Machine Learning

Artificial intelligence (AI) and machine learning are poised to revolutionize communication by enabling more personalized, efficient, and intelligent interactions. AI-powered virtual assistants, such as Siri, Alexa, and Google Assistant, are becoming increasingly sophisticated, offering a wide range of services and support.

Machine learning algorithms are also enhancing communication platforms by improving language translation, content recommendation, and user experience. These technologies have the potential to bridge language barriers, facilitate global collaboration, and create more intuitive and responsive communication systems.

Virtual and Augmented Reality

Virtual reality (VR) and augmented reality (AR) technologies are transforming the way people experience and interact with digital content. VR provides immersive, 3D environments that can be used for communication, entertainment, education, and training. AR overlays digital information onto the physical world, enhancing real-time interactions and providing new ways to access and share information.

These technologies have the potential to revolutionize remote communication, creating more engaging and interactive experiences. For example, VR and AR can enable virtual meetings, conferences, and social gatherings, allowing people to interact in realistic and dynamic virtual environments.

Quantum Communication

Quantum communication, based on the principles of quantum mechanics, promises to revolutionize the security and speed of information exchange. Quantum key distribution (QKD) enables the creation of unbreakable encryption keys, ensuring secure communication even in the presence of eavesdroppers. The development of quantum networks and quantum Internet could significantly enhance the security and efficiency of data transmission, with far-reaching implications for various industries.

The rise of communication technologies, from the telegraph and telephone to the Internet, has had a profound and lasting impact on human society. These innovations have transformed the way people connect, share information, and interact with one another, breaking down barriers of time and distance and creating a highly interconnected global community.

The telegraph revolutionized written communication, enabling rapid information exchange over long distances. The telephone brought voice communication to the forefront, making it possible for people to have real-time conversations across the globe. The Internet has fundamentally transformed every aspect of modern life, providing unprecedented access to information, communication, and commerce.

As we look to the future, emerging technologies such as 5G, AI, VR, and quantum communication promise to further revolutionize communication, creating new opportunities and challenges. The continued evolution of communication technology will shape the way people interact, collaborate, and share knowledge, driving innovation and progress in the 21st century and beyond.

The journey of communication technology is a testament to human ingenuity and the enduring quest to connect and understand one another. From the telegraph to the Internet, each innovation has brought the world closer together, fostering greater connectivity, collaboration, and understanding. As we continue to explore and develop new ways to communicate, the legacy of these groundbreaking technologies will continue to inspire and guide us toward a more connected and informed future.

6. MEDICAL MARVELS

Breakthroughs in Health and Medicine

The story of human progress is profoundly intertwined with the history of medicine. From ancient herbal remedies to modern genetic engineering, the field of medicine has consistently evolved, driven by curiosity, necessity, and innovation. Breakthroughs in health and medicine have not only extended human lifespan but have also vastly improved the quality of life. This chapter explores some of the most significant medical advancements, highlighting the key figures, discoveries, and technologies that have revolutionized healthcare and shaped the course of human history.

Ancient Foundations of Medicine

The origins of medicine can be traced back to ancient civilizations, where early practitioners developed rudimentary methods for treating illness and injury. These early medical practices laid the groundwork for future advancements.

Ancient Egypt and Greece

In ancient Egypt, medicine was a blend of practical treatments and spiritual rituals. The Edwin Smith Papyrus, dating back to around 1600 BCE, is one of the oldest known medical texts and provides insights into Egyptian surgical practices and treatments. Egyptian physicians, such as Imhotep, who was later deified, are credited with early surgical techniques and the use of various medicinal herbs.

Ancient Greece made significant contributions to the field of medicine, particularly through the work of Hippocrates, often regarded as the "Father of Medicine." Hippocrates emphasized the importance of observation and diagnosis and introduced the concept of the four humors (blood, phlegm, yellow bile, and black bile) as the basis for understanding disease. The Hippocratic Corpus, a collection of medical texts attributed to Hippocrates and his followers, laid the foundation for medical ethics and practice.

Roman Medicine

The Romans built upon Greek medical knowledge, emphasizing public health and hygiene. They constructed aqueducts, sewers, and public baths to improve sanitation and prevent disease. Galen, a prominent Roman physician, made significant contributions to anatomy and physiology through his dissections and writings. His theories dominated Western medicine for centuries.

The Renaissance and Scientific Revolution

The Renaissance and Scientific Revolution marked a period of renewed interest in science and the natural world, leading to significant advancements in medicine.

Andreas Vesalius and Human Anatomy

Andreas Vesalius, a Flemish anatomist, revolutionized the study of human anatomy through his meticulous dissections and observations. His work "De humani corporis fabrica" ("On the Fabric of the Human Body"), published in 1543, provided detailed and accurate anatomical drawings, challenging many of the inaccuracies found in the works of Galen. Vesalius' emphasis on direct observation and dissection as methods for studying the human body laid the groundwork for modern anatomy.

William Harvey and the Circulatory System

William Harvey, an English physician, made groundbreaking discoveries in the field of physiology with his work on the circulatory system. In 1628, Harvey published "De Motu Cordis" ("On the Motion of the Heart and Blood"), in which he described the continuous circulation of blood throughout the body, pumped by the heart. This discovery challenged the prevailing Galenic model and provided a more accurate understanding of cardiovascular physiology.

Antonie van Leeuwenhoek and Microscopy

Antonie van Leeuwenhoek, a Dutch scientist, is often referred to as the "Father of Microbiology." Using simple microscopes of his own design, Leeuwenhoek was the first to observe and describe microorganisms, including bacteria and protozoa. His discoveries, published in the late 17th century, opened up a new world of microscopic life and laid the foundation for the field of microbiology.

The 19th Century: An Era of Transformation

The 19th century was a period of profound transformation in medicine, marked by significant advancements in medical science, surgery, and public health.

Germ Theory and Microbiology

The development of germ theory fundamentally changed the understanding of disease and led to major advancements in medical science and public health.

Louis Pasteur

Louis Pasteur, a French chemist and microbiologist, made groundbreaking contributions to the understanding of microbial life and disease. In the mid-19th century, Pasteur conducted experiments that disproved the theory of spontaneous generation and demonstrated that microorganisms were responsible for fermentation and spoilage. His work laid the foundation for germ theory, which posited that specific microorganisms cause specific diseases.

Pasteur's research led to the development of pasteurization, a process for killing harmful bacteria in food and drink, and the creation of vaccines for diseases such as rabies and anthrax. His contributions to microbiology and immunology had a profound impact on public health and the prevention of infectious diseases.

Robert Koch

Robert Koch, a German physician and microbiologist, further advanced germ theory by identifying the specific causative agents of several diseases. In the late 19th century, Koch developed a series of techniques for isolating and identifying bacteria, known as Koch's postulates. These methods allowed him to identify the bacteria responsible for tuberculosis, cholera, and anthrax.

Koch's discoveries provided definitive evidence for germ theory and revolutionized the diagnosis and treatment of infectious diseases. His work earned him the Nobel Prize in Physiology or Medicine in 1905 and solidified his legacy as one of the founders of modern bacteriology.

Advances in Surgery and Anesthesia

The 19th century also saw significant advancements in surgical techniques and the development of anesthesia, transforming the field of surgery.

Joseph Lister and Antiseptic Surgery

Joseph Lister, a British surgeon, revolutionized surgical practice by introducing antiseptic techniques to prevent infection. Building on the work of Louis Pasteur, Lister hypothesized that microorganisms in the air could contaminate surgical wounds and cause infection.

In the 1860s, Lister began using carbolic acid (phenol) to sterilize surgical instruments, dressings, and the operative field. His antiseptic methods dramatically reduced the incidence of postoperative infections and mortality rates. Lister's work laid the foundation for modern aseptic techniques and transformed surgery into a safer and more effective practice.

Anesthesia and Pain Management

Before the development of anesthesia, surgery was an excruciating and often traumatic experience for patients. The introduction of anesthesia in the mid-19th century revolutionized surgery by allowing patients to undergo procedures without pain.

In 1846, William T. G. Morton, an American dentist, successfully demonstrated the use of ether as an anesthetic during a surgical procedure at Massachusetts General Hospital. This landmark event marked the beginning of modern anesthesia.

Following Morton's demonstration, other anesthetics, such as chloroform and nitrous oxide, were developed and adopted for surgical use. The advent of anesthesia enabled more complex and invasive surgical procedures, significantly advancing the field of surgery and improving patient outcomes.

Public Health and Epidemiology

The 19th century also saw significant advancements in public health and the study of disease patterns, leading to improved sanitation and the prevention of epidemics.

John Snow and Cholera

John Snow, an English physician, is often regarded as one of the founders of modern epidemiology. In the mid-19th century, Snow conducted groundbreaking research on cholera, a deadly infectious disease that caused several devastating epidemics in London.

During a cholera outbreak in 1854, Snow mapped the cases of cholera in the Soho district of London and identified a contaminated public water pump as the source of the infection. His removal of the pump handle led to a decline in cases, providing strong evidence for the waterborne transmission of cholera.

Snow's work demonstrated the importance of epidemiology in understanding and controlling infectious diseases. His contributions laid the groundwork for modern public health measures, such as sanitation, clean water supply, and disease surveillance.

The 20th Century: Modern Medicine Emerges

The 20th century was a period of unprecedented advancements in medicine, marked by the discovery of antibiotics, the development of vaccines, and significant progress in medical technology and research.

The Discovery of Antibiotics

The discovery of antibiotics revolutionized the treatment of bacterial infections and saved countless lives.

Alexander Fleming and Penicillin

In 1928, Alexander Fleming, a Scottish bacteriologist, made a serendipitous discovery that would change the course of medicine. While working at St. Mary's Hospital in London, Fleming observed that a mold (Penicillium notatum) growing on a petri dish had killed the surrounding bacteria. He identified the mold's active substance as penicillin, the first true antibiotic.

Penicillin was initially difficult to produce in large quantities, but during World War II, a team of scientists, including Howard Florey and Ernst Boris Chain, developed methods for mass production. The widespread use of penicillin during the war saved countless lives and marked the beginning of the antibiotic era.

Penicillin's success spurred the discovery and development of other antibiotics, revolutionizing the treatment of bacterial infections and significantly reducing mortality rates.

Vaccines and Immunization

The development of vaccines has been one of the most effective public health measures for preventing infectious diseases.

Edward Jenner and Smallpox

The concept of vaccination dates back to the late 18th century, when Edward Jenner, an English physician, developed the first successful smallpox vaccine. In 1796, Jenner inoculated a young boy with material from a cowpox lesion and later demonstrated that the boy was immune to smallpox.

Jenner's work laid the foundation for the development of vaccines and the practice of immunization. The global eradication of smallpox, declared by the World Health Organization (WHO) in 1980, stands as one of the greatest achievements in public health.

Jonas Salk and Polio

The 20th century saw the development of several important vaccines, including the polio vaccine. Polio, a crippling and potentially deadly disease, caused widespread fear and outbreaks throughout the early 20th century.

In the 1950s, Jonas Salk, an American virologist, developed the first effective polio vaccine using inactivated poliovirus. The vaccine was introduced in 1955 and led to a dramatic decline in polio cases. Albert Sabin later developed an oral polio vaccine using live attenuated virus, further aiding global immunization efforts.

The success of the polio vaccines has brought the world close to eradicating the disease, with only a few cases reported annually in a handful of countries.

Medical Imaging and Diagnostic Technologies

Advancements in medical imaging and diagnostic technologies have revolutionized the ability to diagnose and treat diseases.

X-Rays and Radiology

The discovery of X-rays by Wilhelm Conrad Roentgen in 1895 revolutionized medical imaging. X-rays allowed for the non-invasive visualization of the internal structures of the body, enabling the diagnosis of fractures, infections, and tumors.

The development of radiology as a medical specialty expanded the use of X-rays and led to the creation of other imaging modalities, such as computed tomography (CT) scans and magnetic resonance imaging (MRI). These technologies provide detailed and accurate images of the body's organs and tissues, aiding in the diagnosis and treatment of a wide range of conditions.

Ultrasound and Medical Sonography

Ultrasound technology, which uses high-frequency sound waves to create images of internal organs and tissues, has become an essential tool in medical diagnostics. Developed in the mid-20th century, ultrasound is widely used in obstetrics, cardiology, and other medical fields.

Ultrasound imaging is non-invasive, safe, and provides real-time images, making it invaluable for monitoring fetal development, assessing cardiac function, and guiding minimally invasive procedures.

Advances in Surgery and Transplantation

The 20th century saw significant advancements in surgical techniques and the development of organ transplantation.

Heart Surgery and Cardiopulmonary Bypass

The development of cardiopulmonary bypass technology in the 1950s revolutionized heart surgery by allowing surgeons to perform complex procedures on a still and bloodless heart. This technology made it possible to perform surgeries such as coronary artery bypass grafting (CABG) and heart valve replacement, significantly improving outcomes for patients with heart disease.

Organ Transplantation

Organ transplantation represents one of the most remarkable achievements in modern medicine. The first successful kidney transplant was performed by Joseph Murray and his team in 1954. This breakthrough was followed by successful transplants of other organs, including the liver, heart, and lungs.

Advancements in immunosuppressive drugs, such as cyclosporine, have improved the success rates of organ transplants by preventing rejection. Organ transplantation has saved countless lives and provided new hope for patients with end-stage organ failure.

The 21st Century: The Age of Precision Medicine and Biotechnology

The 21st century has ushered in an era of precision medicine and biotechnology, characterized by personalized treatments and cutting-edge innovations.

Genomics and Personalized Medicine

The completion of the Human Genome Project in 2003 marked a milestone in understanding the genetic basis of human health and disease. The project mapped the entire human genome, providing insights into the genetic factors that contribute to various conditions.

Personalized medicine, also known as precision medicine, leverages genomic information to tailor treatments to individual patients. By understanding a patient's genetic makeup, healthcare providers can develop targeted therapies that are more effective and have fewer side effects.

Immunotherapy and Cancer Treatment

Immunotherapy represents a groundbreaking approach to cancer treatment, harnessing the body's immune system to target and destroy cancer cells. Advances in immunotherapy, such as checkpoint inhibitors and CAR-T cell therapy, have shown remarkable success in treating certain types of cancer.

Checkpoint inhibitors, such as pembrolizumab and nivolumab, work by blocking proteins that prevent immune cells from attacking cancer cells. CAR-T cell therapy involves modifying a patient's T cells to recognize and attack cancer cells. These therapies have shown promising results in treating cancers such as melanoma, lung cancer, and leukemia.

Regenerative Medicine and Stem Cell Therapy

Regenerative medicine and stem cell therapy hold the potential to revolutionize the treatment of a wide range of conditions by repairing or replacing damaged tissues and organs.

Stem cells, which have the ability to differentiate into various cell types, offer promising avenues for treating conditions such as spinal cord injuries, Parkinson's disease, and heart disease. Research in regenerative medicine aims to develop therapies that can restore function and improve outcomes for patients with chronic and degenerative conditions.

Advances in Medical Technology

The rapid advancement of medical technology continues to drive innovation in healthcare.

Artificial Intelligence and Machine Learning

Artificial intelligence (AI) and machine learning are transforming medical diagnostics and treatment planning. AI algorithms can analyze vast amounts of medical data, including imaging studies, genetic information, and electronic health records, to identify patterns and make accurate diagnoses.

Machine learning is also being used to develop predictive models for disease progression and treatment response, enabling personalized and evidence-based care.

Robotics and Minimally Invasive Surgery

Robotic surgery has revolutionized minimally invasive procedures by providing surgeons with enhanced precision, control, and visualization. Robotic systems, such as the da Vinci Surgical System, enable surgeons to perform complex surgeries through small incisions, reducing recovery time and improving patient outcomes.

Minimally invasive surgery, including laparoscopic and endoscopic techniques, has become the standard of care for many procedures, offering benefits such as reduced pain, shorter hospital stays, and faster recovery.

The history of medicine is a testament to human ingenuity, perseverance, and the relentless pursuit of knowledge. From ancient herbal remedies to cutting-edge biotechnology, medical advancements have transformed the way we understand and treat disease, improving the quality of life for countless individuals.

The breakthroughs in health and medicine explored in this chapter highlight the remarkable progress that has been made over the centuries. These innovations have not only extended human lifespan but have also provided new hope for patients facing a wide range of conditions.

As we continue to push the boundaries of medical science and technology, the future holds even greater promise. Advances in genomics, immunotherapy, regenerative medicine, and artificial intelligence are poised to revolutionize healthcare, offering new possibilities for diagnosis, treatment, and prevention.

The journey of medical innovation is far from over. As we build upon the foundations of past discoveries, we remain committed to advancing the frontiers of medicine, driven by the goal of improving health and well-being for all.

7. FROM THE EARTH TO THE STARS

Space Exploration and Astronomy

The desire to understand the cosmos and our place within it is as old as humanity itself. From ancient stargazers who mapped the night sky to modern scientists who send probes to the far reaches of the solar system, the quest for knowledge about space has driven some of the most profound and transformative innovations in science and technology. This chapter explores the milestones of space exploration and astronomy, highlighting key discoveries, technological advancements, and the enduring impact of our journey from the Earth to the stars.

Ancient Astronomy: The Beginnings of Celestial Observation

Long before the advent of modern technology, ancient civilizations were captivated by the night sky. They observed the motions of the stars, planets, and celestial phenomena, using their knowledge for practical purposes such as navigation, agriculture, and the marking of time.

Ancient Civilizations and the Night Sky

Many ancient cultures developed sophisticated systems of astronomy. The Babylonians, Egyptians, Greeks, Chinese, and Mayans, among others, made significant contributions to early celestial observation.

The Babylonians

The Babylonians, who lived in Mesopotamia around 1800 BCE, were among the first to systematically record the movements of celestial bodies. They created detailed star catalogs, identified the zodiac, and developed mathematical models to predict celestial events like eclipses. Their observations laid the groundwork for future astronomical studies.

The Egyptians

The ancient Egyptians used their knowledge of astronomy to align their monuments with celestial bodies. The Great Pyramid of Giza, for example, is aligned with the cardinal points and certain stars. The Egyptians also developed a calendar based on the heliacal rising of the star Sirius, which was crucial for predicting the annual flooding of the Nile River.

The Greeks

Greek astronomers made significant strides in understanding the cosmos. Pythagoras, Ptolemy, and Aristotle developed geocentric models of the universe, placing the Earth at the center. However, it was Aristarchus of Samos who first proposed a heliocentric model, suggesting that the Earth orbits the Sun. Although not widely accepted at the time, Aristarchus' idea would later be vindicated by the work of Copernicus.

The Contributions of Islamic Scholars

During the Islamic Golden Age (8th to 14th centuries), Muslim scholars made substantial contributions to astronomy. They preserved and expanded upon the knowledge of earlier civilizations, translating Greek and Roman texts into Arabic and conducting their own observations.

Al-Battani

Al-Battani, a 9th-century astronomer, made precise measurements of the solar year and the orbits of the planets. His work on trigonometry and astronomical calculations influenced later European astronomers.

Al-Sufi

Al-Sufi's "Book of Fixed Stars," written in the 10th century, provided detailed descriptions and illustrations of the constellations. He also made accurate measurements of the brightness and positions of stars, some of which are still in use today.

Ibn al-Haytham

Ibn al-Haytham, also known as Alhazen, made significant contributions to optics and the scientific method. His work on the nature of light and vision laid the foundations for modern optical astronomy.

The Renaissance and the Scientific Revolution

The Renaissance and the Scientific Revolution marked a turning point in the study of astronomy. During this period, groundbreaking discoveries and the development of new instruments transformed our understanding of the universe.

Nicolaus Copernicus and the Heliocentric Model

In the early 16th century, Nicolaus Copernicus, a Polish astronomer, proposed a heliocentric model of the solar system. In his seminal work, "De revolutionibus orbium coelestium" ("On the Revolutions of the Celestial Spheres"), published in 1543, Copernicus argued that the Earth and other planets orbit the Sun. This revolutionary idea challenged the geocentric model that had dominated Western thought for over a millennium.

Although Copernicus' heliocentric theory was initially met with resistance, it provided a more accurate framework for understanding the motions of celestial bodies. His work laid the foundation for future astronomical discoveries and set the stage for the Scientific Revolution.

Galileo Galilei and the Telescope

In the early 17th century, Galileo Galilei, an Italian astronomer and physicist, made significant contributions to astronomy through his use of the telescope. Galileo constructed his own telescopes and used them to make a series of groundbreaking observations.

In 1610, Galileo published "Sidereus Nuncius" ("Starry Messenger"), in which he described his discoveries of the moons of Jupiter, the phases of Venus, and the rugged surface of the Moon. These observations provided strong evidence for the heliocentric model and challenged the prevailing Aristotelian view of a perfect, unchanging celestial realm.

Galileo's work laid the groundwork for modern observational astronomy and demonstrated the power of the telescope as a scientific instrument.

Johannes Kepler and the Laws of Planetary Motion

Johannes Kepler, a German mathematician and astronomer, built upon the work of Copernicus and Galileo to develop his laws of planetary motion. Kepler's meticulous analysis of astronomical data, particularly the observations of Mars by Tycho Brahe, led him to formulate three fundamental laws that described the motion of planets around the Sun.

Kepler's laws, published in "Astronomia Nova" (1609) and "Harmonices Mundi" (1619), demonstrated that planetary orbits are elliptical rather than circular, that planets sweep out equal areas in equal times, and that there is a precise mathematical relationship between the orbital period of a planet and its distance from the Sun. These laws provided a more accurate and predictive model of the solar system, further validating the heliocentric theory.

Isaac Newton and Universal Gravitation

Isaac Newton, an English mathematician, physicist, and astronomer, is perhaps the most celebrated figure of the Scientific Revolution. His work "Philosophiæ Naturalis Principia Mathematica" ("Mathematical Principles of Natural Philosophy"), published in 1687, laid the foundation for classical mechanics and revolutionized our understanding of the natural world.

In the "Principia," Newton formulated the three laws of motion, which describe the relationship between the motion of an object and the forces acting upon it. These laws provided a comprehensive framework for understanding the behavior of objects in motion and formed the basis for Newtonian physics.

Newton's law of universal gravitation, also presented in the "Principia," posited that every mass in the universe exerts an attractive force on every other mass. This force, known as gravity, explained the motions of celestial bodies and unified the physics of the heavens and the Earth under a single theoretical framework.

The 20th Century: The Age of Space Exploration

The 20th century marked the beginning of a new era in astronomy and space exploration. The development of rocketry and space technology enabled humanity to venture beyond the confines of Earth and explore the cosmos firsthand.

The Space Race

The Space Race, a period of intense competition between the United States and the Soviet Union, was a defining feature of the mid-20th century. This rivalry spurred rapid advancements in space technology and led to some of the most significant achievements in the history of space exploration.

Sputnik and the Dawn of the Space Age

On October 4, 1957, the Soviet Union launched Sputnik 1, the world's first artificial satellite. This historic event marked the beginning of the Space Age and demonstrated the potential of space technology. Sputnik 1's success shocked the world and prompted the United States to accelerate its own space program.

The First Human in Space

The Soviet Union achieved another milestone on April 12, 1961, when Yuri Gagarin became the first human to travel into space. Gagarin's spacecraft, Vostok 1, completed a single orbit of Earth, making him an international hero and solidifying the Soviet Union's lead in the Space Race.

The Apollo Program and the Moon Landing

In response to the Soviet achievements, the United States launched the Apollo program with the goal of landing a man on the Moon. On July 20, 1969, NASA's Apollo 11 mission achieved this historic feat when astronauts Neil Armstrong and Edwin "Buzz" Aldrin set foot on the lunar surface. Armstrong's famous words, "That's one small step for [a] man, one giant leap for mankind," captured the significance of the moment.

The Apollo program, which included a total of six manned lunar landings, provided valuable scientific data about the Moon and demonstrated the capabilities of human spaceflight. The success of Apollo 11 remains one of the most iconic achievements in the history of space exploration.

Robotic Exploration of the Solar System

While human spaceflight captured the public's imagination, robotic missions have played a crucial role in exploring the solar system and beyond. These missions have provided detailed observations and data about distant planets, moons, and other celestial bodies.

The Mariner and Voyager Missions

NASA's Mariner program, which began in the 1960s, included a series of robotic missions to explore the inner planets of the solar system. Mariner 2 became the first spacecraft to fly by Venus in 1962, and Mariner 4 conducted the first successful flyby of Mars in 1965, sending back the first close-up images of the Martian surface.

The Voyager program, launched in 1977, included two spacecraft, Voyager 1 and Voyager 2, designed to explore the outer planets. The Voyagers provided detailed images and data about Jupiter, Saturn, Uranus, and Neptune, transforming our understanding of these distant worlds. Voyager 1 has since become the first human-made object to enter interstellar space, continuing its journey beyond the solar system.

The Mars Rovers

Mars has been a primary target for robotic exploration due to its potential for past or present life. NASA's Mars rover missions, including Spirit, Opportunity, Curiosity, and Perseverance, have conducted detailed investigations of the Martian surface. These rovers have analyzed soil and rock samples, searched for signs of water, and provided valuable insights into the planet's geology and climate.

Perseverance, which landed on Mars in 2021, is equipped with advanced scientific instruments and is tasked with searching for evidence of ancient microbial life. The mission also includes the Ingenuity helicopter, the first aircraft to achieve powered flight on another planet.

The Hubble Space Telescope

The Hubble Space Telescope, launched in 1990, has revolutionized our understanding of the universe by providing stunning images and detailed observations of distant galaxies, nebulae, and other celestial phenomena. Hubble's observations have contributed to numerous scientific discoveries, including the determination of the universe's expansion rate and the detection of exoplanets.

International Collaboration in Space Exploration

Space exploration has increasingly become a collaborative endeavor, with international partnerships playing a crucial role in advancing our understanding of the cosmos.

The International Space Station

The International Space Station (ISS), a joint project involving NASA, Roscosmos, ESA, JAXA, and CSA, serves as a symbol of international cooperation in space. Launched in 1998, the ISS orbits Earth and serves as a laboratory for scientific research and technology development. The station has hosted astronauts from around the world and has conducted experiments in fields such as biology, physics, astronomy, and Earth science.

The European Space Agency

The European Space Agency (ESA) has been a key player in space exploration, contributing to numerous scientific missions and collaborations. ESA's Rosetta mission, launched in 2004, achieved a historic first by successfully landing the Philae probe on the comet 67P/Churyumov-Gerasimenko in 2014. The mission provided valuable data about the composition and behavior of comets.

ESA has also collaborated with NASA on missions such as the Hubble Space Telescope and the Mars Express orbiter, which has been studying the Martian surface and atmosphere since 2003.

China's Space Program

China has emerged as a major player in space exploration, achieving significant milestones in recent years. The China National Space Administration (CNSA) has launched a series of successful missions, including the Chang'e lunar exploration program and the Tianwen Mars mission.

In 2019, the Chang'e 4 mission achieved a historic first by landing a rover on the far side of the Moon. The Tianwen-1 mission, launched in 2020, successfully placed an orbiter around Mars and landed the Zhurong rover on the Martian surface in 2021.

The Future of Space Exploration

The future of space exploration promises to be as exciting and transformative as its past, with ambitious missions and groundbreaking technologies on the horizon.

Return to the Moon

NASA's Artemis program aims to return humans to the Moon by the mid-2020s. The program plans to land the first woman and the next man on the lunar surface, establish a sustainable presence on the Moon, and pave the way for future missions to Mars. International partners and private companies are expected to play a significant role in achieving these goals.

Mars and Beyond

Human missions to Mars represent the next major milestone in space exploration. NASA, SpaceX, and other organizations are developing the technologies and plans needed to send astronauts to the Red Planet. These missions will aim to explore the potential for past or present life on Mars, as well as test technologies for future human exploration of the solar system.

The Search for Extraterrestrial Life

The search for extraterrestrial life continues to be a major focus of space exploration. Missions such as the James Webb Space Telescope, scheduled for launch in 2021, will provide unprecedented capabilities for studying the atmospheres of exoplanets and searching for signs of habitability.

Robotic missions to the icy moons of Jupiter and Saturn, such as Europa Clipper and Dragonfly, aim to explore the potential for subsurface oceans and the conditions for life beyond Earth.

Space Tourism and Commercial Spaceflight

The rise of commercial spaceflight companies, such as SpaceX, Blue Origin, and Virgin Galactic, is opening new possibilities for space tourism and private missions. These companies are developing reusable rockets and spacecraft, making space travel more affordable and accessible.

Space tourism promises to offer unique experiences for individuals, while commercial spaceflight has the potential to support scientific research, satellite deployment, and future space exploration missions.

The journey from the Earth to the stars has been marked by remarkable achievements, driven by humanity's innate curiosity and desire to explore the unknown. From ancient astronomers who first mapped the night sky to modern scientists who send probes to distant planets, the quest for knowledge about the cosmos has led to some of the most profound innovations in science and technology.

Space exploration and astronomy have not only expanded our understanding of the universe but have also inspired generations of scientists, engineers, and dreamers. The discoveries and advancements made in these fields have transformed our world, providing new insights into the nature of reality and our place within it.

As we look to the future, the spirit of exploration and discovery continues to drive us forward. With ambitious missions and cutting-edge technologies on the horizon, the next chapter in the story of space exploration promises to be as exciting and transformative as the ones that came before. The journey from the Earth to the stars is far from over, and the possibilities for discovery and innovation are boundless.

8. DIGITAL REVOLUTION

The Birth and Evolution of Computing

The digital revolution is one of the most transformative periods in human history, fundamentally altering how we live, work, and communicate. The rise of computing has driven this revolution, with profound impacts on every aspect of modern society. This chapter explores the birth and evolution of computing, tracing its origins from early mechanical devices to the sophisticated digital technologies that underpin today's information age.

Early Computing: Mechanical Foundations

The journey of computing began long before the advent of digital electronics, with early mechanical devices designed to perform calculations and solve complex problems.

The Abacus

The abacus, one of the earliest known calculating tools, dates back to ancient civilizations such as Sumeria, Egypt, China, and Greece. This simple device, consisting of beads on rods, enabled users to perform basic arithmetic operations such as addition, subtraction, multiplication, and division. The abacus remained in use for thousands of years and laid the groundwork for more advanced computational tools.

The Antikythera Mechanism

Discovered in a shipwreck off the coast of the Greek island of Antikythera in 1901, the Antikythera mechanism is an ancient Greek analog computer dating back to around 100 BCE. This intricate device used a complex system of gears to predict astronomical positions and eclipses. The Antikythera mechanism is a testament to the advanced engineering and mathematical knowledge of the ancient Greeks.

John Napier and Logarithms

In the early 17th century, Scottish mathematician John Napier invented logarithms, a mathematical concept that simplified complex calculations. Napier's invention of logarithms led to the creation of logarithmic tables and slide rules, which became essential tools for engineers and scientists for centuries.

Blaise Pascal and the Pascaline

Blaise Pascal, a French mathematician and philosopher, invented one of the first mechanical calculators, the Pascaline, in 1642. The Pascaline used a series of gears and dials to perform addition and subtraction. While the Pascaline had limited commercial success, it represented a significant step forward in the development of mechanical computing devices.

Gottfried Wilhelm Leibniz and the Stepped Reckoner

German polymath Gottfried Wilhelm Leibniz improved upon Pascal's design with his Stepped Reckoner, built in the 1670s. The Stepped Reckoner could perform multiplication and division in addition to addition and subtraction. Leibniz also developed the binary number system, which would later become the foundation of digital computing.

Charles Babbage and the Analytical Engine

Charles Babbage, an English mathematician and inventor, is often considered the "father of the computer" for his design of the Analytical Engine. Conceived in the 1830s, the Analytical Engine was a mechanical general-purpose computer that could perform any calculation through the use of punched cards for input and output. Although Babbage never completed a working model, his designs laid the theoretical foundation for modern computers.

The Dawn of Electronic Computing

The early 20th century saw significant advancements in electronics and the development of the first electronic computers, which marked the beginning of the digital age.

Alan Turing and the Turing Machine

Alan Turing, a British mathematician and logician, made groundbreaking contributions to the field of computer science with his concept of the Turing machine. Introduced in 1936, the Turing machine was a theoretical model that defined the principles of computation. Turing's work provided a formal foundation for the design and analysis of algorithms and is considered one of the cornerstones of theoretical computer science.

The Colossus and Codebreaking

During World War II, the British government developed the Colossus, the world's first programmable electronic digital computer, to break encrypted German messages. Designed by engineer Tommy Flowers and operational by 1944, the Colossus played a crucial role in deciphering the Lorenz cipher used by the German High Command. The success of the Colossus demonstrated the potential of electronic computing for complex problem-solving.

The ENIAC and Early Electronic Computers

In the United States, the Electronic Numerical Integrator and Computer (ENIAC) was developed during World War II by John W. Mauchly and J. Presper Eckert at the University of Pennsylvania. Completed in 1945, the ENIAC was the first general-purpose electronic digital computer. It used vacuum tubes to perform calculations and could be programmed to solve a wide range of mathematical problems.

The ENIAC's success led to the development of other early electronic computers, such as the EDVAC (Electronic Discrete Variable Automatic Computer) and the UNIVAC I (Universal Automatic Computer). These early computers laid the groundwork for the rapid advancement of computing technology in the following decades.

The Transition to Transistors and Integrated Circuits

The invention of the transistor and the development of integrated circuits revolutionized computing, making computers smaller, faster, and more reliable.

The Invention of the Transistor

In 1947, John Bardeen, Walter Brattain, and William Shockley at Bell Labs invented the transistor, a semiconductor device that could amplify and switch electronic signals. The transistor replaced bulky and unreliable vacuum tubes, enabling the development of smaller and more efficient electronic devices. The invention of the transistor earned the trio the Nobel Prize in Physics in 1956 and paved the way for the miniaturization of computers.

The Development of Integrated Circuits

In the late 1950s, Jack Kilby at Texas Instruments and Robert Noyce at Fairchild Semiconductor independently developed the integrated circuit (IC), a small chip containing multiple transistors and other electronic components. The integrated circuit allowed for the mass production of complex electronic circuits on a single chip, significantly reducing the size and cost of electronic devices.

The introduction of integrated circuits led to the development of microprocessors, the central processing units (CPUs) of modern computers. Microprocessors enabled the creation of personal computers and revolutionized the electronics industry.

The Rise of Personal Computing

The 1970s and 1980s saw the emergence of personal computers (PCs), bringing computing power to individuals and transforming the way people interact with technology.

The Altair 8800 and the Birth of the PC

The Altair 8800, released in 1975 by Micro Instrumentation and Telemetry Systems (MITS), is often considered the first commercially successful personal computer. Based on the Intel 8080 microprocessor, the Altair 8800 was sold as a kit that hobbyists could assemble. Its success sparked the personal computing revolution and inspired the creation of numerous other PC models.

Apple and the Macintosh

In 1976, Steve Jobs and Steve Wozniak founded Apple Computer, Inc. and introduced the Apple I, a single-board computer designed for hobbyists. The following year, Apple released the Apple II, one of the first successful mass-produced personal computers. The Apple II featured a color display, expandable memory, and a variety of software applications, making it popular with both home users and businesses.

In 1984, Apple launched the Macintosh, a groundbreaking personal computer with a graphical user interface (GUI) and a mouse. The Macintosh's intuitive design and user-friendly interface set a new standard for personal computing and influenced the development of future operating systems.

IBM and the PC Revolution

IBM entered the personal computer market in 1981 with the release of the IBM Personal Computer (IBM PC). The IBM PC was based on the Intel 8088 microprocessor and ran the Microsoft Disk Operating System (MS-DOS). Its open architecture and widespread adoption by businesses and consumers established the IBM PC as the industry standard.

The success of the IBM PC led to the proliferation of "IBM-compatible" PCs, which were manufactured by various companies and ran the same software. This compatibility standardization helped drive the rapid growth of the personal computer market.

Microsoft and the Software Industry

Microsoft, founded by Bill Gates and Paul Allen in 1975, played a pivotal role in the personal computing revolution. The company's MS-DOS operating system became the standard for IBM-compatible PCs, and its software products, such as Microsoft Word and Microsoft Excel, became essential tools for personal and business use.

In 1985, Microsoft released Windows, an operating system with a graphical user interface that built upon the success of MS-DOS. Windows quickly became the dominant operating system for personal computers, solidifying Microsoft's position as a leading software company.

The Internet and the Information Age

The development of the Internet and the World Wide Web transformed computing and communication, ushering in the Information Age and connecting people and information on a global scale.

The Origins of the Internet

The origins of the Internet can be traced back to the 1960s, when the United States Department of Defense initiated research on a communication network that could withstand potential disruptions caused by a nuclear attack. This project, known as ARPANET (Advanced Research Projects Agency Network), laid the foundation for the modern Internet.

ARPANET was designed to connect computers at various research institutions, allowing them to share information and resources. The first successful message was sent over ARPANET in 1969, marking the beginning of computer networking.

The Development of TCP/IP and the Expansion of the Internet

One of the key developments in the evolution of the Internet was the creation of the Transmission Control Protocol/Internet Protocol (TCP/IP). This set of communication protocols, developed by Vint Cerf and Robert Kahn in the 1970s, provided a standardized way for different networks to communicate with each other. TCP/IP became the foundational protocol for the Internet, enabling the seamless exchange of data between diverse computer systems.

Throughout the 1980s and 1990s, the Internet expanded rapidly. The introduction of personal computers, modems, and networking technologies made it possible for individuals and businesses to connect to the Internet. The development of the World Wide Web by Tim Berners-Lee in 1989 further accelerated the growth of the Internet by providing a user-friendly interface for accessing and sharing information.

The World Wide Web and the Information Age

The World Wide Web (WWW), often simply referred to as the Web, revolutionized the way people interact with the Internet. Berners-Lee's creation, which included the concepts of hypertext and hyperlinks, allowed users to navigate the Internet through web pages and links. The first web browser, Mosaic, was released in 1993, making the Web accessible to a broader audience.

The Web transformed the Internet into a vast repository of information, enabling users to access a wide range of content, from academic research to entertainment. It also facilitated the development of e-commerce, social media, and online communication platforms, fundamentally changing the way people conduct business, socialize, and consume media.

The Dot-Com Boom and the Rise of Tech Giants

The late 1990s and early 2000s saw the rise of the dot-com boom, a period of rapid growth and investment in Internet-based companies. Many startups emerged, offering innovative online services and e-commerce solutions. While the dot-com bubble eventually burst, leading to the collapse of many companies, it also paved the way for the growth of tech giants such as Amazon, Google, and eBay.

These companies, along with others like Facebook and Apple, have become dominant players in the technology industry, shaping the way we interact with information and each other. Their products and services have revolutionized e-commerce, search, social networking, and mobile computing, making them integral to modern life.

The Era of Mobile Computing and the Cloud

The 21st century has witnessed the rise of mobile computing and cloud technology, further transforming the landscape of computing and connectivity.

The Smartphone Revolution

The introduction of smartphones revolutionized personal computing by integrating powerful computing capabilities with mobile communication. The release of Apple's iPhone in 2007 marked a turning point, setting new standards for design, functionality, and user experience. The iPhone's success spurred the development of competing smartphones, such as those running Google's Android operating system.

Smartphones have become ubiquitous, providing users with access to the Internet, apps, and services on the go. They have transformed how we communicate, access information, and interact with the digital world, making mobile computing an essential part of everyday life.

The Rise of Cloud Computing

Cloud computing has revolutionized how data and applications are stored, accessed, and managed. Instead of relying on local hardware and software, cloud computing allows users to access resources over the Internet. This shift has enabled greater flexibility, scalability, and collaboration.

Services such as Amazon Web Services (AWS), Microsoft Azure, and Google Cloud Platform provide cloud-based infrastructure, platforms, and software, supporting a wide range of applications from data storage to machine learning. Cloud computing has become the backbone of modern IT infrastructure, powering everything from streaming services to enterprise applications.

The Internet of Things (IoT)

The Internet of Things (IoT) represents the next frontier in computing, connecting everyday objects to the Internet and enabling them to collect and exchange data. IoT devices, such as smart home appliances, wearable fitness trackers, and industrial sensors, are transforming how we interact with the physical world.

IoT technology is driving innovation across various sectors, including healthcare, transportation, and manufacturing. It has the potential to improve efficiency, enhance user experiences, and enable new business models by harnessing the power of data and connectivity.

The Future of Computing

As we look to the future, emerging technologies promise to further revolutionize computing and its applications, pushing the boundaries of what is possible.

Artificial Intelligence and Machine Learning

Artificial intelligence (AI) and machine learning are transforming computing by enabling machines to learn from data and make decisions. AI-powered technologies, such as natural language processing, computer vision, and autonomous systems, are being integrated into a wide range of applications, from healthcare to finance to transportation.

Advancements in AI are driving the development of intelligent assistants, predictive analytics, and automated processes, reshaping industries and improving efficiency. As AI continues to evolve, it has the potential to revolutionize every aspect of society.

Quantum Computing

Quantum computing represents a paradigm shift in computing technology, leveraging the principles of quantum mechanics to perform calculations that are impossible for classical computers. Quantum computers use qubits, which can exist in multiple states simultaneously, enabling them to solve complex problems at unprecedented speeds.

While still in the experimental stage, quantum computing holds promise for solving challenges in cryptography, materials science, drug discovery, and optimization. Researchers and companies are investing in the development of quantum hardware and algorithms, aiming to unlock the transformative potential of this technology.

Edge Computing

Edge computing is an emerging paradigm that brings computation and data storage closer to the sources of data, such as IoT devices and sensors. By processing data at the edge of the network, edge computing reduces latency, improves response times, and reduces the load on centralized cloud servers.

Edge computing is particularly important for applications that require real-time processing and low latency, such as autonomous vehicles, industrial automation, and smart cities. It complements cloud computing by enabling distributed and scalable computing architectures.

5G and Beyond

The rollout of 5G networks is set to revolutionize connectivity by providing faster speeds, lower latency, and greater capacity. 5G technology will enable new applications and services, from augmented reality and virtual reality to smart infrastructure and telemedicine.

Beyond 5G, researchers are exploring next-generation wireless technologies, such as 6G, which promise even greater performance and capabilities. These advancements will drive the continued evolution of computing and connectivity, supporting the growing demands of a digital society.

The digital revolution, driven by the birth and evolution of computing, has fundamentally transformed the world. From early mechanical devices to modern digital technologies, the advancements in computing have reshaped every aspect of human life, enabling unprecedented levels of connectivity, innovation, and progress.

As we continue to push the boundaries of technology, the future of computing holds immense promise. Emerging technologies such as AI, quantum computing, edge computing, and 5G are poised to revolutionize industries, improve quality of life, and address some of the most pressing challenges facing humanity.

The journey of computing is a testament to human ingenuity, curiosity, and the relentless pursuit of knowledge. As we move forward into the digital age, the innovations and discoveries of the past and present will continue to inspire and guide us toward a future of limitless possibilities.

9. ENERGY INNOVATIONS

From Steam Engines to Renewable Power

Energy is the lifeblood of modern civilization. The ability to harness and utilize various forms of energy has driven technological progress, economic growth, and societal development. From the invention of the steam engine to the rise of renewable energy sources, innovations in energy technology have transformed the world in profound ways. This chapter explores the evolution of energy innovations, tracing the journey from the early days of steam power to the cutting-edge renewable technologies that promise to shape the future.

The Steam Engine: Powering the Industrial Revolution

The advent of the steam engine in the 18th century marked a pivotal moment in the history of energy technology. It played a central role in powering the Industrial Revolution, transforming industries, transportation, and society at large.

Early Steam Engines

The concept of using steam to perform work dates back to ancient times. Hero of Alexandria, a Greek engineer, described a simple steam-powered device called the aeolipile in the first century CE. However, it wasn't until the early modern period that practical steam engines were developed.

Thomas Newcomen and the Atmospheric Engine

In 1712, Thomas Newcomen, an English blacksmith and inventor, developed the first practical steam engine, known as the atmospheric engine.

Newcomen's engine was primarily used for pumping water out of coal mines. It worked by creating a vacuum to draw water up through a cylinder using atmospheric pressure. Although relatively inefficient, Newcomen's engine marked the beginning of the steam-powered industrial age.

James Watt and the Improved Steam Engine

The steam engine saw significant improvements with the work of James Watt, a Scottish engineer and inventor. In the 1760s, Watt developed a separate condenser, which greatly increased the efficiency of the steam engine by reducing energy losses. Watt's engine also featured a rotary motion mechanism, making it more versatile for various industrial applications.

Watt's innovations revolutionized the steam engine, making it a reliable and powerful source of mechanical energy. His partnership with entrepreneur Matthew Boulton led to the widespread adoption of steam engines in factories, mills, and mines, driving the rapid industrialization of Britain and other countries.

The Impact of Steam Power

The steam engine had a transformative impact on various sectors of the economy and society:

Industrial Manufacturing

Steam power enabled the mechanization of manufacturing processes, increasing productivity and efficiency. Factories equipped with steam engines could operate machinery continuously, leading to mass production and the growth of industries such as textiles, metallurgy, and engineering.

Transportation

Steam engines revolutionized transportation with the development of steamships and locomotives. In 1807, Robert Fulton's steamboat, the Clermont, successfully navigated the Hudson River, demonstrating the potential of steam-powered water transportation. The introduction of steam locomotives, pioneered by George Stephenson's Rocket in 1829, revolutionized land transportation by enabling faster and more reliable movement of goods and people over long distances.

Urbanization and Society

The Industrial Revolution, powered by steam engines, led to rapid urbanization as people moved from rural areas to cities in search of work. This shift had profound social and economic implications, contributing to the rise of a new industrial working class and the expansion of urban infrastructure.

The Age of Electricity: Lighting Up the World

The discovery and harnessing of electricity in the 19th and 20th centuries marked another major milestone in the history of energy innovation. Electricity transformed everyday life, enabling new technologies and powering the modern world.

Early Discoveries and Developments

The study of electricity dates back to ancient times, but significant scientific advancements occurred in the 18th and 19th centuries.

Benjamin Franklin and Electricity

Benjamin Franklin, an American polymath, conducted pioneering experiments with electricity in the mid-18th century. His famous kite experiment in 1752 demonstrated that lightning is a form of electricity. Franklin's work laid the groundwork for the understanding of electrical phenomena and the development of practical applications.

Alessandro Volta and the Electric Battery

In 1800, Alessandro Volta, an Italian physicist, invented the electric battery, known as the voltaic pile. Volta's battery provided a continuous and reliable source of electrical current, enabling further experimentation and discoveries in electromagnetism and electrochemistry.

The Invention of the Electric Generator

The invention of the electric generator, also known as the dynamo, was a key development in the harnessing of electrical energy.

Michael Faraday and Electromagnetic Induction

Michael Faraday, an English scientist, discovered the principle of electromagnetic induction in 1831. Faraday demonstrated that a changing magnetic field could induce an electric current in a conductor. This discovery led to the development of the first electric generators, which converted mechanical energy into electrical energy.

Thomas Edison and Direct Current (DC) Power

Thomas Edison, an American inventor and entrepreneur, played a pivotal role in the development of practical electric power systems. In the late 19th century, Edison developed the first commercially viable incandescent light bulb and established the first electric power generation and distribution system in New York City in 1882. Edison's system used direct current (DC) electricity, which had limitations in terms of transmission distance.

Nikola Tesla and Alternating Current (AC) Power

Nikola Tesla, a Serbian-American inventor and electrical engineer, made significant contributions to the development of alternating current (AC) power systems. Tesla's AC system, which allowed for efficient transmission of electricity over long distances, competed with Edison's DC system in what became known as the "War of Currents." Tesla's partnership with industrialist George Westinghouse led to the widespread adoption of AC power, which became the standard for electric power distribution.

The Electrification of Society

The widespread adoption of electricity had a profound impact on society, transforming daily life, industry, and technology:

Lighting and Appliances

Electric lighting, powered by incandescent bulbs and later fluorescent and LED technology, revolutionized indoor and outdoor illumination. The availability of electric power also led to the development of a wide range of household appliances, such as electric stoves, refrigerators, washing machines, and vacuum cleaners, which improved the quality of life and increased convenience.

Communication

Electricity enabled the development of new communication technologies, including the telegraph, telephone, and radio. These innovations revolutionized long-distance communication, connecting people and information across the globe.

Industry and Automation

Electric power facilitated the development of automated machinery and assembly lines, increasing industrial productivity and efficiency. The electrification of factories and transportation systems contributed to economic growth and technological advancement.

The Rise of Fossil Fuels: Oil, Gas, and Coal

The 20th century saw the rise of fossil fuels—oil, natural gas, and coal—as the dominant sources of energy. These fuels powered industrial growth, transportation, and the development of modern infrastructure.

The Oil Industry

The discovery and exploitation of petroleum resources transformed the energy landscape, driving economic growth and technological innovation.

The Birth of the Oil Industry

The modern oil industry began in the mid-19th century with the drilling of the first successful oil well by Edwin Drake in Titusville, Pennsylvania, in 1859. The discovery of oil in large quantities led to the rapid growth of the oil industry, with major oil fields developed in the United States, Russia, the Middle East, and other regions.

The Development of the Internal Combustion Engine

The invention and development of the internal combustion engine revolutionized transportation and industry. German engineers Nikolaus Otto and Karl Benz made significant contributions to the development of gasoline-powered engines in the late 19th century. The widespread adoption of internal combustion engines powered automobiles, trucks, ships, and airplanes, transforming global transportation.

The Role of Oil Companies

Major oil companies, such as Standard Oil (founded by John D. Rockefeller), Royal Dutch Shell, and British Petroleum (BP), played a crucial role in the exploration, production, and distribution of oil. These companies became dominant players in the global energy market, shaping the geopolitics of oil.

Natural Gas

Natural gas emerged as a valuable energy resource, used for heating, electricity generation, and industrial processes.

Early Uses of Natural Gas

Natural gas has been used for heating and lighting since ancient times. However, its widespread commercial use began in the 19th century with the development of pipelines and distribution networks.

The Expansion of Natural Gas Infrastructure

The 20th century saw significant investments in natural gas infrastructure, including pipelines, storage facilities, and liquefied natural gas (LNG) terminals. Advances in drilling technology, such as hydraulic fracturing (fracking) and horizontal drilling, increased the availability of natural gas and expanded its use as a clean and efficient energy source.

Coal: The Backbone of Industrialization

Coal played a central role in powering the Industrial Revolution and remained a dominant energy source throughout the 20th century.

The Role of Coal in Industrialization

Coal was the primary fuel for steam engines, furnaces, and power plants during the Industrial Revolution. Its abundance and energy density made it a key resource for industrial growth and economic development.

Environmental and Health Impacts

The extensive use of coal has had significant environmental and health impacts, including air pollution, greenhouse gas emissions, and respiratory diseases. The negative effects of coal combustion have led to efforts to reduce coal consumption and transition to cleaner energy sources.

The Transition to Renewable Energy

The late 20th and early 21st centuries have seen a growing recognition of the need for sustainable and environmentally friendly energy sources. The transition to renewable energy has become a central focus of energy innovation and policy.

The Rise of Solar Power

Solar power has emerged as a key player in the renewable energy landscape, harnessing the energy of the sun to generate electricity and heat.

Photovoltaic (PV) Technology

Photovoltaic (PV) technology, which converts sunlight directly into electricity using semiconductor materials, has seen significant advancements in efficiency and cost reduction. The development of silicon-based solar cells in the mid-20th century paved the way for the widespread adoption of solar power.

Concentrated Solar Power (CSP)

Concentrated Solar Power (CSP) technology uses mirrors or lenses to concentrate sunlight onto a small area, generating heat that can be used to produce electricity. CSP systems can include thermal energy storage, allowing for electricity generation even when the sun is not shining.

Solar Energy Adoption

The adoption of solar energy has been driven by falling costs, government incentives, and technological advancements. Solar power is now used for residential, commercial, and utility-scale applications, contributing to the global transition to renewable energy.

Wind Energy

Wind energy has become one of the fastest-growing renewable energy sources, harnessing the power of wind to generate electricity.

Development of Wind Turbines

Modern wind turbines, which convert kinetic energy from the wind into electrical energy, have seen significant improvements in design, efficiency, and capacity. The development of larger and more efficient turbines has enabled the expansion of wind farms on land and offshore.

Wind Energy Deployment

Wind energy has seen rapid growth in recent decades, with significant investments in wind farm development and grid integration. Countries such as Denmark, Germany, and the United States have become leaders in wind energy production, contributing to the diversification of the energy mix.

Hydropower

Hydropower, one of the oldest forms of renewable energy, continues to play a crucial role in global electricity generation.

Development of Hydroelectric Dams

Hydroelectric dams, which use the potential energy of stored water to generate electricity, have been developed around the world. Large-scale projects, such as the Hoover Dam in the United States and the Three Gorges Dam in China, have provided substantial amounts of renewable energy.

Small-Scale and Run-of-River Hydropower

In addition to large-scale hydroelectric dams, small-scale and run-of-river hydropower systems have been developed to generate electricity without significant environmental impact. These systems harness the energy of flowing water in rivers and streams, providing renewable energy to remote and rural areas.

Geothermal Energy

Geothermal energy harnesses the heat from within the Earth to generate electricity and provide direct heating.

Development of Geothermal Power Plants

Geothermal power plants use steam or hot water from geothermal reservoirs to drive turbines and generate electricity. Countries with significant geothermal resources, such as Iceland, the Philippines, and the United States, have developed geothermal power plants to harness this renewable energy source.

Direct Use of Geothermal Energy

Geothermal energy is also used directly for heating buildings, greenhouses, and industrial processes. Geothermal heat pumps, which transfer heat between the ground and buildings, provide efficient and sustainable heating and cooling solutions.

Bioenergy

Bioenergy, derived from organic materials such as biomass, biogas, and biofuels, is a versatile and renewable energy source.

Biomass Energy

Biomass energy is produced by burning organic materials, such as wood, agricultural residues, and waste, to generate heat and electricity. Biomass power plants and combined heat and power (CHP) systems provide renewable energy while reducing waste and emissions.

Biogas and Anaerobic Digestion

Biogas, produced through the anaerobic digestion of organic materials, is a renewable source of methane that can be used for electricity generation, heating, and transportation. Anaerobic digestion systems, which process agricultural, industrial, and municipal waste, provide renewable energy and contribute to waste management.

Biofuels

Biofuels, such as ethanol and biodiesel, are renewable alternatives to fossil fuels for transportation. Biofuels are produced from feedstocks such as corn, sugarcane, and vegetable oils. The development of advanced biofuels from non-food sources, such as algae and cellulosic biomass, holds promise for sustainable transportation.

The Future of Energy: Innovations and Challenges

As the world continues to transition to renewable energy, ongoing innovations and challenges will shape the future of energy.

Energy Storage

Energy storage technologies, such as batteries, pumped hydro, and thermal storage, are critical for integrating renewable energy into the grid and ensuring reliable power supply. Advances in battery technology, such as lithium-ion and solid-state batteries, are driving the development of energy storage solutions for both grid and transportation applications.

Smart Grids

Smart grids use digital technologies to enhance the efficiency, reliability, and flexibility of the electricity grid. Smart grids enable better integration of renewable energy, demand response, and distributed generation, contributing to a more sustainable and resilient energy system.

Energy Efficiency

Improving energy efficiency is a key strategy for reducing energy consumption and emissions. Innovations in building design, industrial processes, and transportation can significantly enhance energy efficiency and contribute to sustainable development.

Hydrogen Economy

Hydrogen is a versatile energy carrier that can be produced from various sources, including renewable electricity. The development of a hydrogen economy, which includes hydrogen production, storage, and utilization, holds promise for decarbonizing sectors such as transportation, industry, and heating.

Policy and Regulation

Effective policy and regulation are essential for supporting the transition to renewable energy and addressing the challenges of climate change. Governments and international organizations play a crucial role in setting targets, providing incentives, and ensuring the deployment of clean energy technologies.

The history of energy innovations is a testament to human ingenuity and the relentless pursuit of progress. From the early steam engines that powered the Industrial Revolution to the cutting-edge renewable technologies of today, each breakthrough has transformed the way we live and interact with the world.

As we face the challenges of climate change and resource depletion, the transition to sustainable and renewable energy is more important than ever. Innovations in energy technology will continue to drive this transition, shaping a future where clean, reliable, and affordable energy is accessible to all.

The journey of energy innovation is far from over. As we build on the achievements of the past and embrace the opportunities of the future, we remain committed to advancing the frontiers of energy technology and creating a sustainable and prosperous world for generations to come.

10. MATERIALS OF TOMORROW

Advancements in Chemistry and Materials Science

The evolution of human civilization has always been closely linked to the materials we have been able to develop and manipulate. From the Stone Age through the Bronze and Iron Ages to the modern era of advanced composites and nanotechnology, the materials we use have defined the tools, structures, and technologies that shape our world. This chapter explores the remarkable advancements in chemistry and materials science that have paved the way for the development of the materials of tomorrow, transforming industries and improving the quality of life.

The Foundations of Materials Science

Materials science, an interdisciplinary field that combines elements of chemistry, physics, and engineering, focuses on understanding the properties of materials and developing new materials with specific characteristics. The roots of materials science can be traced back to the early study of metallurgy and the development of synthetic materials in the 19th and 20th centuries.

Metallurgy and the Development of Alloys

Metallurgy, the science of extracting and refining metals from ores, has been practiced for thousands of years. The development of alloys, mixtures of two or more metals, marked significant milestones in the history of materials.

Bronze Age

The Bronze Age, which began around 3300 BCE, saw the widespread use of bronze, an alloy of copper and tin. Bronze tools and weapons were stronger and more durable than those made of stone, leading to advancements in agriculture, warfare, and craftsmanship.

Iron Age

The Iron Age, starting around 1200 BCE, introduced the use of iron and steel. Iron tools and weapons were even stronger and more versatile than their bronze counterparts. The ability to produce and work with iron revolutionized many aspects of life, from construction to transportation.

The Advent of Synthetic Materials

The 19th and 20th centuries witnessed the development of synthetic materials, which expanded the range of available materials beyond those found in nature.

Polymers and Plastics

The discovery of polymers, long-chain molecules made up of repeating units, led to the development of plastics, a class of materials with a wide range of properties. In 1907, Leo Baekeland invented Bakelite, the first synthetic plastic, which was used in electrical insulators, radio casings, and a variety of other products.

The development of other polymers, such as polyethylene, polypropylene, and polyvinyl chloride (PVC), revolutionized industries from packaging to construction. Plastics offered advantages such as lightweight, durability, and resistance to corrosion, making them indispensable in modern life.

Synthetic Fibers

The invention of synthetic fibers, such as nylon, polyester, and Kevlar, transformed the textile industry. Nylon, developed by Wallace Carothers at DuPont in 1935, was the first commercially successful synthetic fiber and found widespread use in clothing, ropes, and other applications. Kevlar, also developed by DuPont, is known for its high tensile strength and is used in bulletproof vests, helmets, and other protective gear.

Modern Advances in Materials Science

The late 20th and early 21st centuries have seen rapid advancements in materials science, driven by developments in chemistry, physics, and engineering. These advancements have led to the creation of materials with unprecedented properties and capabilities.

Nanotechnology

Nanotechnology, the manipulation of matter on an atomic and molecular scale, has opened up new possibilities for materials science. Nanomaterials, which have structures with dimensions on the nanoscale (typically less than 100 nanometers), exhibit unique properties that differ from those of bulk materials.

Carbon Nanotubes

Carbon nanotubes, cylindrical structures made of carbon atoms arranged in a hexagonal lattice, are among the most well-known nanomaterials. Discovered in 1991 by Sumio Iijima, carbon nanotubes are incredibly strong, lightweight, and conductive. They have potential applications in fields ranging from electronics to materials reinforcement to drug delivery.

Graphene

Graphene, a single layer of carbon atoms arranged in a two-dimensional honeycomb lattice, has garnered significant attention since its isolation in 2004 by Andre Geim and Konstantin Novoselov. Graphene is known for its exceptional strength, electrical conductivity, and thermal conductivity. Its potential applications include flexible electronics, high-performance batteries, and advanced composites.

Advanced Composites

Composite materials, made from two or more constituent materials with different properties, have been used for centuries. However, recent advancements have led to the development of advanced composites with superior performance characteristics.

Fiber-Reinforced Polymers

Fiber-reinforced polymers (FRPs), which consist of a polymer matrix reinforced with fibers such as carbon or glass, are widely used in aerospace, automotive, and construction industries. Carbon fiber-reinforced polymers

(CFRPs) offer high strength-to-weight ratios and are used in applications where weight savings are critical, such as aircraft and high-performance sports equipment.

Metal Matrix Composites

Metal matrix composites (MMCs), which consist of a metal matrix reinforced with ceramic or other metal particles, offer enhanced mechanical properties and wear resistance. MMCs are used in applications such as automotive engine components, cutting tools, and aerospace structures.

Smart Materials

Smart materials, also known as responsive or adaptive materials, can change their properties in response to external stimuli such as temperature, pressure, or electric fields.

Shape Memory Alloys

Shape memory alloys (SMAs) are materials that can return to their original shape after being deformed when exposed to a specific temperature. Nickel-titanium (NiTi) alloys, also known as Nitinol, are among the most well-known SMAs. They are used in applications such as medical devices, actuators, and aerospace components.

Piezoelectric Materials

Piezoelectric materials generate an electric charge in response to mechanical stress and can also deform when subjected to an electric field. These materials are used in sensors, actuators, and energy harvesting devices.

High-Temperature Superconductors

Superconductors are materials that can conduct electricity without resistance below a certain critical temperature. The discovery of high-temperature superconductors, which operate at relatively higher temperatures compared to traditional superconductors, has opened up new possibilities for applications in energy transmission, magnetic levitation, and medical imaging.

Biomaterials

Biomaterials, materials designed for use in medical applications, have seen significant advancements in recent years. These materials are used in implants, prosthetics, and tissue engineering.

Biocompatible Polymers

Biocompatible polymers, such as polylactic acid (PLA) and polycaprolactone (PCL), are used in medical devices and drug delivery systems. These polymers can degrade safely in the body, making them suitable for temporary implants and controlled release of medications.

Tissue Engineering

Tissue engineering involves the use of biomaterials, cells, and biochemical factors to create or repair tissues and organs. Advances in 3D printing and regenerative medicine have enabled the development of tissue scaffolds and organoids, which have the potential to revolutionize transplantation and personalized medicine.

Applications and Impact of Advanced Materials

The development of advanced materials has had a profound impact on various industries and aspects of daily life. These materials have enabled technological advancements, improved performance, and enhanced sustainability.

Aerospace and Aviation

The aerospace and aviation industries have benefited significantly from the development of advanced materials.

Lightweight Composites

The use of lightweight composites, such as CFRPs, has led to more fuel-efficient aircraft with improved performance. The Boeing 787 Dreamliner and the Airbus A350 are examples of commercial aircraft that extensively use composite materials to reduce weight and enhance fuel efficiency.

High-Temperature Materials

High-temperature materials, such as ceramic matrix composites (CMCs) and superalloys, are used in jet engines and spacecraft to withstand extreme conditions. These materials enable higher operating temperatures, improving engine efficiency and performance.

Automotive Industry

Advanced materials have transformed the automotive industry by improving vehicle performance, safety, and fuel efficiency.

Lightweight Materials

The use of lightweight materials, such as aluminum alloys, high-strength steels, and composites, has led to the development of lighter and more fuel-efficient vehicles. These materials help reduce greenhouse gas emissions and improve overall vehicle performance.

Safety Materials

Advanced materials, such as high-strength steels and energy-absorbing polymers, are used in automotive safety components, including crumple zones and airbags. These materials enhance passenger protection and improve vehicle crashworthiness.

Electronics and Communication

The electronics and communication industries have been revolutionized by the development of advanced materials.

Semiconductors

Semiconductor materials, such as silicon, gallium arsenide, and indium phosphide, are the foundation of modern electronics. Advances in semiconductor technology have enabled the development of smaller, faster, and more efficient electronic devices, from microprocessors to smartphones.

Conductive Polymers

Conductive polymers, such as polyaniline and polythiophene, are used in flexible electronics, organic light-emitting diodes (OLEDs), and wearable devices. These materials offer advantages such as flexibility, lightweight, and ease of processing.

Energy and Sustainability

Advanced materials play a critical role in addressing energy and sustainability challenges.

Photovoltaic Materials

Advances in photovoltaic materials have led to the development of more efficient and cost-effective solar panels. Materials such as perovskites and thin-film semiconductors have the potential to significantly improve solar energy conversion efficiency.

Energy Storage

Materials science has driven advancements in energy storage technologies, such as batteries and supercapacitors. Lithium-ion batteries, widely used in electric vehicles and portable electronics, have benefited from developments in electrode materials and electrolytes. Emerging energy storage technologies, such as solid-state batteries and redox flow batteries, hold promise for grid-scale energy storage.

Environmental Remediation

Advanced materials are used in environmental remediation to remove pollutants from air, water, and soil. Materials such as activated carbon, zeolites, and photocatalysts are employed in filtration systems, adsorption processes, and catalytic degradation of contaminants.

Healthcare and Medicine

The healthcare and medical industries have seen significant advancements due to the development of advanced materials.

Implants and Prosthetics

Biocompatible materials, such as titanium alloys and medical-grade polymers, are used in implants and prosthetics. These materials offer advantages such as biocompatibility, strength, and durability, improving patient outcomes and quality of life.

Drug Delivery Systems

Advanced materials are used in drug delivery systems to improve the efficacy and safety of medications. Nanoparticles, liposomes, and hydrogels are employed to deliver drugs in a controlled and targeted manner, reducing side effects and enhancing therapeutic outcomes.

Construction and Infrastructure

Advanced materials have transformed the construction and infrastructure industries, enabling the development of more durable, sustainable, and efficient buildings and structures.

High-Performance Concrete

High-performance concrete, reinforced with fibers and admixtures, offers improved strength, durability, and resistance to environmental factors. This material is used in infrastructure projects such as bridges, tunnels, and high-rise buildings.

Sustainable Building Materials

Sustainable building materials, such as recycled aggregates, bamboo, and bio-based composites, are used in green building projects to reduce environmental impact and enhance energy efficiency. These materials contribute to sustainable development and resource conservation.

Future Directions in Materials Science

The future of materials science holds immense potential for further advancements and innovations. Emerging trends and research areas are likely to shape the development of materials in the coming decades.

Nanomaterials and Nanotechnology

Nanomaterials and nanotechnology will continue to play a significant role in the development of advanced materials. Research in areas such as carbon nanomaterials, quantum dots, and nanocomposites will lead to new applications and improved performance in electronics, energy, and healthcare.

2D Materials

Two-dimensional (2D) materials, such as graphene, transition metal dichalcogenides (TMDs), and hexagonal boron nitride, offer unique properties and potential for a wide range of applications. Advances in the synthesis, characterization, and integration of 2D materials will drive innovations in flexible electronics, sensors, and photonics.

Bioinspired and Biomimetic Materials

Bioinspired and biomimetic materials, designed to mimic the structure and function of natural materials, hold promise for a variety of applications. Research in areas such as self-healing materials, bioadhesives, and synthetic biomaterials will lead to new solutions in healthcare, construction, and environmental protection.

Additive Manufacturing

Additive manufacturing, also known as 3D printing, is revolutionizing the production of materials and components. Advances in additive manufacturing technologies and materials, such as metal powders, polymers, and ceramics, will enable the creation of complex, high-performance structures with applications in aerospace, healthcare, and beyond.

Quantum Materials

Quantum materials, which exhibit exotic properties due to quantum mechanical effects, are an emerging research area with potential for transformative applications. Materials such as topological insulators, quantum spin liquids, and high-temperature superconductors are being explored for use in quantum computing, spintronics, and advanced electronics.

The advancements in chemistry and materials science have had a profound impact on the development of new materials with exceptional properties and capabilities. From the early days of metallurgy to the cutting-edge research in nanotechnology and quantum materials, the field of materials science has continuously pushed the boundaries of what is possible.

The materials of tomorrow will play a critical role in addressing the challenges of the 21st century, from sustainable energy and environmental protection to healthcare and advanced technology. As we continue to explore and innovate, the future of materials science holds immense promise for creating a better, more sustainable, and technologically advanced world.

11. ARTIFICIAL INTELLIGENCE

The New Frontier of Technology

Artificial Intelligence (AI) is rapidly becoming one of the most transformative technologies of the 21st century. As we stand on the brink of a new era in technology, AI is poised to revolutionize industries, enhance human capabilities, and redefine the boundaries of what machines can achieve. This chapter explores the development, applications, and implications of AI, tracing its journey from early theoretical concepts to its current state as a powerful tool driving innovation and change.

The Origins of Artificial Intelligence

The concept of artificial intelligence has its roots in ancient mythology and early philosophical musings about the nature of intelligence and the possibility of creating artificial beings. However, the formal study and development of AI began in the 20th century, driven by advancements in mathematics, computer science, and cognitive science.

Early Foundations

The theoretical foundations of AI can be traced back to the work of several key figures and concepts in the early 20th century.

Alan Turing and the Turing Machine

Alan Turing, a British mathematician and logician, is often considered one of the founding figures of AI. In 1936, Turing introduced the concept of the

Turing machine, a theoretical device capable of performing any computation given the right algorithm and sufficient time. Turing's work laid the groundwork for the development of digital computers and the formalization of algorithms.

In 1950, Turing published a seminal paper titled "Computing Machinery and Intelligence," in which he proposed the famous Turing Test. The test, designed to evaluate a machine's ability to exhibit intelligent behavior indistinguishable from that of a human, remains a benchmark for AI research.

Early Cybernetics and Control Systems

The field of cybernetics, which emerged in the 1940s and 1950s, focused on the study of communication and control in living organisms and machines. Pioneers such as Norbert Wiener and W. Ross Ashby explored concepts of feedback, adaptation, and self-organization, which would later influence AI research.

The Birth of AI Research

The formal establishment of AI as a field of study is often attributed to the Dartmouth Conference in 1956. Organized by John McCarthy, Marvin Minsky, Nathaniel Rochester, and Claude Shannon, the conference brought together leading researchers to explore the possibility of creating machines capable of intelligent behavior.

Early AI Programs

In the years following the Dartmouth Conference, researchers developed some of the first AI programs, which demonstrated rudimentary forms of machine intelligence.

- **Logic Theorist**: Developed by Allen Newell and Herbert A. Simon in 1955, the Logic Theorist was one of the first AI programs. It was designed to prove mathematical theorems by mimicking human problem-solving processes.
- **General Problem Solver (GPS)**: Also developed by Newell and Simon, the GPS was an early attempt to create a general-purpose problem-solving machine. It used heuristics and means-ends analysis to solve a wide range of problems.

- **ELIZA**: Created by Joseph Weizenbaum in the mid-1960s, ELIZA was an early natural language processing program that simulated a conversation with a human by using pattern matching and substitution rules. ELIZA's famous "DOCTOR" script mimicked a Rogerian psychotherapist, highlighting the potential for AI in human-computer interaction.

The Evolution of AI: From Symbolic AI to Machine Learning

The development of AI has been marked by several distinct phases, each characterized by different approaches and methodologies. Two major paradigms in AI research are symbolic AI and machine learning.

Symbolic AI

Symbolic AI, also known as "good old-fashioned AI" (GOFAI), dominated AI research from the 1950s to the 1980s. This approach focuses on the manipulation of symbols and the use of formal logic to represent knowledge and solve problems.

Expert Systems

One of the most successful applications of symbolic AI was the development of expert systems in the 1970s and 1980s. Expert systems are computer programs that emulate the decision-making abilities of human experts in specific domains. They use a knowledge base of facts and rules, along with an inference engine, to solve complex problems.

- **MYCIN**: Developed in the 1970s, MYCIN was an expert system designed to diagnose bacterial infections and recommend treatments. It used a rule-based approach to model the knowledge of medical experts and achieved impressive diagnostic accuracy.
- **DENDRAL**: Another early expert system, DENDRAL, was developed to analyze chemical compounds and infer molecular structures from mass spectrometry data. DENDRAL demonstrated the potential of AI to assist in scientific discovery and research.

The Rise of Machine Learning

In the late 20th century, the focus of AI research began to shift from symbolic approaches to machine learning, a paradigm that emphasizes the use of statistical methods and data-driven algorithms to enable machines to learn from experience.

Artificial Neural Networks

Artificial neural networks, inspired by the structure and function of the human brain, became a central focus of machine learning research. Early work on neural networks, such as the Perceptron developed by Frank Rosenblatt in the 1950s, laid the foundation for future advancements.

- **Backpropagation**: In the 1980s, the development of the backpropagation algorithm, which enables the training of multi-layer neural networks, revitalized interest in neural networks. Researchers such as Geoffrey Hinton, David Rumelhart, and Ronald J. Williams demonstrated the potential of deep learning for complex pattern recognition tasks.

Support Vector Machines and Kernel Methods

Support vector machines (SVMs) and kernel methods, developed in the 1990s, became popular techniques for classification and regression tasks. These methods use mathematical optimization to find the best decision boundary between different classes of data.

Ensemble Methods

Ensemble methods, such as boosting and bagging, combine multiple machine learning models to improve predictive performance. Techniques like AdaBoost and Random Forests demonstrated the power of ensemble learning for a wide range of applications.

The AI Renaissance: Deep Learning and Modern AI

The 21st century has witnessed a resurgence of interest in AI, driven by advances in computational power, the availability of large datasets, and breakthroughs in deep learning.

The Deep Learning Revolution

Deep learning, a subfield of machine learning that focuses on training deep neural networks, has become the driving force behind many of the most significant advancements in AI.

Convolutional Neural Networks (CNNs)

Convolutional neural networks (CNNs), designed for processing structured grid data such as images, have revolutionized computer vision. The breakthrough performance of CNNs in image recognition tasks, demonstrated by the AlexNet model in the 2012 ImageNet competition, marked a turning point for deep learning.

- **Applications of CNNs**: CNNs are used in a wide range of applications, including image classification, object detection, facial recognition, and medical image analysis. They have enabled significant advancements in fields such as autonomous driving, security, and healthcare.

Recurrent Neural Networks (RNNs) and Transformers

Recurrent neural networks (RNNs) are designed for processing sequential data, such as time series and natural language. Variants of RNNs, such as long short-term memory (LSTM) networks and gated recurrent units (GRUs), have shown success in tasks like language modeling and speech recognition.

Transformers, introduced in 2017 by Vaswani et al., have revolutionized natural language processing (NLP). The self-attention mechanism in transformers enables efficient modeling of long-range dependencies in text, leading to significant improvements in language understanding and generation.

- **Applications of Transformers**: Transformers power state-of-the-art NLP models, such as BERT (Bidirectional Encoder Representations from Transformers), GPT-3 (Generative Pre-trained Transformer 3), and T5 (Text-to-Text Transfer Transformer). These models have achieved breakthroughs in tasks like machine translation, text summarization, sentiment analysis, and conversational AI.

AI in Practice: Applications and Impact

The advancements in AI have led to its widespread adoption across various industries and domains, transforming how we live and work.

Healthcare

AI is revolutionizing healthcare by enhancing diagnostics, personalized medicine, and drug discovery.

- **Medical Imaging**: AI algorithms, particularly CNNs, are used to analyze medical images for the detection and diagnosis of conditions such as cancer, cardiovascular diseases, and neurological disorders. AI-powered imaging tools assist radiologists in identifying abnormalities with high accuracy.
- **Personalized Medicine**: AI enables the analysis of large-scale genomic and clinical data to identify personalized treatment plans for patients. Machine learning models can predict patient responses to therapies and optimize treatment protocols.
- **Drug Discovery**: AI accelerates the drug discovery process by analyzing molecular structures, predicting drug-target interactions, and identifying potential therapeutic candidates. AI-powered platforms help pharmaceutical companies streamline research and development.

Finance

AI is transforming the finance industry by enhancing decision-making, risk management, and customer experience.

- **Algorithmic Trading**: Machine learning algorithms analyze market data and execute trades at high speeds, optimizing trading strategies and maximizing returns. AI-driven trading systems can adapt to changing market conditions and identify patterns that human traders might miss.
- **Fraud Detection**: AI models detect fraudulent activities by analyzing transaction patterns and identifying anomalies. Machine learning algorithms continuously learn from new data to improve the accuracy of fraud detection and prevention.
- **Customer Service**: AI-powered chatbots and virtual assistants provide personalized customer support, handling inquiries and transactions efficiently. Natural language processing enables these systems to understand and respond to customer queries in real time.

Manufacturing

AI is revolutionizing manufacturing by enhancing automation, quality control, and predictive maintenance.

- **Automation**: AI-powered robots and autonomous systems perform complex tasks with precision and efficiency. Machine learning algorithms optimize manufacturing processes, reducing waste and improving productivity.
- **Quality Control**: Computer vision systems powered by AI inspect products for defects and ensure high-quality standards. AI models analyze sensor data to detect deviations and anomalies in real time.
- **Predictive Maintenance**: AI algorithms analyze data from machinery and equipment to predict maintenance needs and prevent failures. Predictive maintenance reduces downtime and extends the lifespan of industrial assets.

Transportation

AI is transforming transportation by enabling autonomous vehicles, optimizing logistics, and improving safety.

- **Autonomous Vehicles**: AI algorithms power self-driving cars, enabling them to perceive their environment, make decisions, and navigate safely. Machine learning models process data from sensors, cameras, and lidar to detect objects and plan routes.
- **Logistics Optimization**: AI optimizes supply chain and logistics operations by analyzing demand patterns, optimizing routes, and managing inventory. Machine learning algorithms improve efficiency and reduce costs in transportation networks.
- **Traffic Management**: AI systems analyze traffic data to optimize traffic flow, reduce congestion, and enhance road safety. Predictive models forecast traffic conditions and enable dynamic traffic control measures.

Retail

AI is transforming the retail industry by enhancing customer experience, inventory management, and marketing.

- **Customer Experience**: AI-powered recommendation systems analyze customer preferences and behavior to provide personalized product recommendations. Virtual shopping assistants and chatbots enhance customer service and support.
- **Inventory Management**: AI algorithms optimize inventory levels by analyzing sales data, demand patterns, and supply chain variables. Machine learning models predict stockouts and overstock situations, improving inventory turnover.
- **Marketing**: AI-driven marketing platforms analyze customer data to optimize advertising campaigns, segment audiences, and personalize marketing messages. Predictive analytics helps retailers identify trends and opportunities.

Ethical and Societal Implications of AI

As AI becomes increasingly integrated into society, it raises important ethical and societal considerations that must be addressed to ensure its responsible and beneficial use.

Bias and Fairness

AI systems can inadvertently perpetuate or amplify biases present in the data they are trained on. Ensuring fairness and mitigating bias in AI models is crucial to prevent discrimination and ensure equitable outcomes.

- **Addressing Bias**: Researchers and practitioners are developing techniques to detect and mitigate bias in AI models, such as fairness-aware machine learning and bias correction algorithms. Transparent and accountable AI development processes are essential for addressing bias.

Privacy and Security

The widespread use of AI involves the collection and analysis of vast amounts of personal data, raising concerns about privacy and security.

- **Data Privacy**: Protecting user privacy requires robust data anonymization, encryption, and secure data handling practices. Regulations such as the General Data Protection Regulation (GDPR) set standards for data protection and privacy.

- **Security**: Ensuring the security of AI systems is critical to prevent malicious attacks and misuse. Techniques such as adversarial training, secure model deployment, and continuous monitoring help protect AI systems from threats.

Transparency and Explainability

AI models, particularly deep learning models, can be complex and difficult to interpret. Ensuring transparency and explainability in AI systems is important for building trust and understanding their decision-making processes.

- **Explainable AI**: Research in explainable AI focuses on developing methods to make AI models more interpretable and understandable to users. Techniques such as feature attribution, model visualization, and rule extraction contribute to explainability.

Impact on Employment

The automation capabilities of AI have the potential to disrupt labor markets and impact employment in various industries.

- **Workforce Transition**: Preparing the workforce for the impact of AI involves reskilling and upskilling initiatives, as well as promoting lifelong learning. Policies and programs that support workers in transitioning to new roles and industries are essential.
- **Human-AI Collaboration**: Emphasizing human-AI collaboration, where AI augments human capabilities rather than replacing jobs, can help mitigate the negative impact on employment. Designing AI systems that complement and enhance human work is key to maximizing benefits.

The Future of AI: Opportunities and Challenges

The future of AI holds immense potential for further advancements and innovations, as well as challenges that need to be addressed.

Advancements in AI Research

Ongoing research in AI aims to push the boundaries of what machines can achieve and explore new frontiers.

- **General AI**: While current AI systems are designed for specific tasks, achieving general AI—machines with human-like cognitive abilities and adaptability—remains a long-term goal. Advances in neural architecture, transfer learning, and meta-learning are steps toward this vision.
- **Human-Centric AI**: Developing AI systems that understand and align with human values, emotions, and intentions is a growing area of research. Human-centric AI aims to create more natural and effective interactions between humans and machines.
- **AI for Science**: AI has the potential to accelerate scientific discovery by analyzing complex data, generating hypotheses, and automating experiments. AI-driven research can contribute to breakthroughs in fields such as physics, biology, and materials science.

Global Collaboration and Governance

Ensuring the responsible development and deployment of AI requires global collaboration and effective governance frameworks.

- **International Cooperation**: Promoting international cooperation and knowledge sharing in AI research and development is essential for addressing global challenges and maximizing benefits. Collaborative initiatives can foster innovation and ethical standards.
- **AI Governance**: Developing comprehensive governance frameworks for AI involves establishing regulations, standards, and best practices that ensure safety, fairness, and accountability. Policymakers, industry leaders, and researchers must work together to create effective governance structures.

Addressing Societal Challenges

AI has the potential to address some of the most pressing societal challenges, from healthcare to climate change.

- **Healthcare and Aging**: AI can improve healthcare delivery, enhance disease prevention, and support healthy aging. Personalized medicine, telehealth, and AI-driven diagnostics are key areas of impact.

- **Climate Change**: AI can contribute to climate change mitigation and adaptation by optimizing energy usage, enhancing environmental monitoring, and supporting sustainable practices. AI-driven solutions can help address the complex challenges of climate change.
- **Education and Inclusion**: AI can enhance educational outcomes by providing personalized learning experiences, supporting educators, and promoting inclusion. AI-powered educational tools can help bridge gaps in access to quality education.

Artificial intelligence represents a new frontier of technology with the potential to transform every aspect of human life. From its early theoretical foundations to the cutting-edge advancements of today, AI has evolved into a powerful tool that drives innovation, enhances capabilities, and addresses complex challenges.

As we navigate the opportunities and implications of AI, it is essential to ensure its responsible and ethical development. By addressing issues of bias, privacy, transparency, and impact on employment, we can harness the potential of AI for the greater good.

The journey of AI is far from over, and the future holds immense promise for further advancements and breakthroughs. As we continue to explore and innovate, AI will remain a central force in shaping the technological landscape and driving progress toward a better and more inclusive world.

12. SUSTAINABLE FUTURES

Innovations in Environmental Science and Technology

As humanity grapples with the pressing challenges of climate change, resource depletion, and environmental degradation, the quest for sustainability has become paramount. Innovations in environmental science and technology are at the forefront of this effort, offering new ways to protect our planet and ensure a sustainable future for generations to come. This chapter explores groundbreaking advancements in environmental science and technology, highlighting their transformative impact on society and the natural world.

The Urgency of Environmental Sustainability

Environmental sustainability is the practice of using natural resources in a way that meets current needs without compromising the ability of future generations to meet their own needs. The urgency of this endeavor is underscored by the accelerating impacts of climate change, loss of biodiversity, and the degradation of ecosystems.

Climate Change and Its Impacts

Climate change, driven primarily by human activities such as burning fossil fuels and deforestation, poses a significant threat to the environment and human societies. The increase in greenhouse gas emissions, particularly carbon dioxide (CO_2) and methane (CH_4), has led to global warming, resulting in rising sea levels, more frequent and severe weather events, and disruptions to ecosystems.

Biodiversity Loss

Biodiversity, the variety of life on Earth, is essential for ecosystem health and resilience. However, human activities such as habitat destruction, pollution, and overexploitation of resources have led to a dramatic decline in biodiversity. The loss of species and ecosystems undermines the natural processes that support life, including pollination, nutrient cycling, and climate regulation.

Resource Depletion

The unsustainable use of natural resources, such as water, minerals, and forests, threatens the availability of these resources for future generations. Overexploitation of resources can lead to environmental degradation, loss of ecosystem services, and increased competition for scarce resources.

Innovations in Environmental Science and Technology

Innovations in environmental science and technology offer new tools and approaches to address these challenges, promoting sustainability and resilience. This section explores key areas of innovation, including renewable energy, sustainable agriculture, waste management, water conservation, and environmental monitoring.

Renewable Energy

The transition to renewable energy sources is a cornerstone of efforts to reduce greenhouse gas emissions and combat climate change. Advances in renewable energy technologies have made it possible to harness the power of the sun, wind, water, and biomass to generate clean and sustainable energy.

Solar Energy

Solar energy is one of the most abundant and widely available renewable energy sources. Advances in photovoltaic (PV) technology have significantly improved the efficiency and affordability of solar panels, making solar energy a viable option for a wide range of applications.

- **Photovoltaic (PV) Technology**: Innovations in PV technology, such as the development of thin-film solar cells and perovskite

solar cells, have increased the efficiency of solar panels while reducing manufacturing costs. These advancements have expanded the adoption of solar energy in residential, commercial, and utility-scale projects.
- **Concentrated Solar Power (CSP)**: Concentrated solar power (CSP) systems use mirrors or lenses to concentrate sunlight onto a small area, generating heat that can be used to produce electricity. CSP systems with thermal energy storage can provide continuous power generation, even when the sun is not shining.

Wind Energy

Wind energy is another key renewable energy source that has seen significant advancements in recent years. Modern wind turbines are capable of generating large amounts of electricity with minimal environmental impact.

- **Offshore Wind**: Offshore wind farms, located in bodies of water, harness the strong and consistent winds over the ocean to generate electricity. Advances in turbine design and foundation technology have made offshore wind a viable and growing sector of the renewable energy market.
- **High-Altitude Wind**: High-altitude wind energy systems, such as airborne wind turbines and kites, aim to capture the stronger and more consistent winds found at higher altitudes. These systems have the potential to generate significant amounts of electricity and expand the geographical range of wind energy.

Bioenergy

Bioenergy, derived from organic materials such as biomass, biogas, and biofuels, offers a renewable and sustainable alternative to fossil fuels.

- **Biomass Energy**: Biomass energy is produced by burning organic materials, such as wood, agricultural residues, and waste, to generate heat and electricity. Advances in biomass conversion technologies, such as gasification and pyrolysis, have improved the efficiency and sustainability of biomass energy production.
- **Biogas and Anaerobic Digestion**: Biogas, produced through the anaerobic digestion of organic materials, is a renewable source

of methane that can be used for electricity generation, heating, and transportation. Anaerobic digestion systems, which process agricultural, industrial, and municipal waste, provide renewable energy and contribute to waste management.
- **Advanced Biofuels**: Advanced biofuels, such as cellulosic ethanol and algae-based fuels, are produced from non-food feedstocks and have the potential to reduce greenhouse gas emissions compared to conventional biofuels. Research and development in advanced biofuels aim to improve their efficiency and scalability.

Sustainable Agriculture

Sustainable agriculture practices aim to produce food in ways that protect the environment, support local economies, and promote social equity. Innovations in agriculture technology and methods are helping to achieve these goals.

Precision Agriculture

Precision agriculture uses technology to optimize crop management and improve agricultural productivity while minimizing environmental impact.

- **Remote Sensing and GIS**: Remote sensing technologies, such as satellites and drones, along with Geographic Information Systems (GIS), provide detailed data on soil conditions, crop health, and environmental factors. This information allows farmers to make informed decisions about planting, irrigation, fertilization, and pest control.
- **IoT and Smart Farming**: The Internet of Things (IoT) enables the integration of sensors, devices, and data analytics in agriculture. Smart farming systems use IoT technology to monitor and manage agricultural operations in real time, improving efficiency and reducing resource use.

Agroecology

Agroecology is an approach to agriculture that emphasizes the integration of ecological principles into farming practices. It promotes biodiversity, soil health, and sustainable land use.

- **Crop Diversification**: Crop diversification, including intercropping and agroforestry, enhances biodiversity and resilience in agricultural systems. Diverse cropping systems can improve soil health, reduce pest and disease pressure, and increase overall productivity.
- **Organic Farming**: Organic farming practices avoid the use of synthetic pesticides and fertilizers, focusing instead on natural inputs and processes. Organic farming supports soil health, reduces chemical runoff, and promotes biodiversity.

Waste Management and Circular Economy

Innovations in waste management and the circular economy aim to reduce waste generation, promote recycling and reuse, and create sustainable materials and products.

Waste-to-Energy

Waste-to-energy technologies convert waste materials into energy, reducing the volume of waste sent to landfills and generating renewable energy.

- **Incineration with Energy Recovery**: Modern waste incineration facilities use advanced combustion technology to generate electricity and heat from waste. These facilities are equipped with pollution control systems to minimize emissions.
- **Anaerobic Digestion**: Anaerobic digestion of organic waste produces biogas, which can be used for electricity generation, heating, or as a vehicle fuel. This process also produces digestate, a nutrient-rich byproduct that can be used as a fertilizer.

Recycling and Material Recovery

Advancements in recycling technology and processes are improving the efficiency and effectiveness of material recovery, reducing the demand for virgin resources.

- **Mechanical Recycling**: Mechanical recycling involves the physical processing of waste materials, such as plastics and metals, to produce new products. Innovations in sorting and

processing technology have improved the quality and yield of recycled materials.
- **Chemical Recycling**: Chemical recycling breaks down complex materials, such as plastics, into their chemical constituents, which can be used to produce new materials. This approach can handle mixed and contaminated waste streams that are challenging for mechanical recycling.

Circular Economy

The circular economy is an economic model that emphasizes the continuous use of resources, minimizing waste and maximizing value. Innovations in product design, materials science, and business models are driving the transition to a circular economy.

- **Product-as-a-Service**: The product-as-a-service model shifts the focus from selling products to providing services, encouraging manufacturers to design durable and repairable products. This model promotes resource efficiency and reduces waste.
- **Industrial Symbiosis**: Industrial symbiosis involves the collaboration of different industries to use each other's byproducts and waste streams as raw materials. This approach reduces waste and resource use while creating economic and environmental benefits.

Water Conservation and Management

Water is a vital resource for life and economic activity, but it is increasingly under pressure from population growth, pollution, and climate change. Innovations in water conservation and management are essential for ensuring sustainable water use.

Water-Efficient Technologies

Advances in water-efficient technologies help reduce water consumption in agriculture, industry, and households.

- **Drip Irrigation**: Drip irrigation systems deliver water directly to the roots of plants, reducing water loss through evaporation and

runoff. This technology is particularly effective in arid regions and for water-intensive crops.
- **Water-Efficient Appliances**: Water-efficient appliances, such as low-flow faucets, toilets, and washing machines, reduce water use in households and businesses. These appliances incorporate technologies that optimize water use without compromising performance.

Water Recycling and Reuse

Water recycling and reuse technologies treat wastewater for safe reuse in various applications, reducing the demand for freshwater resources.

- **Greywater Recycling**: Greywater recycling systems treat wastewater from sinks, showers, and laundry for non-potable uses such as irrigation and toilet flushing. These systems reduce the demand for freshwater and decrease the volume of wastewater discharged into the environment.
- **Industrial Water Reuse**: Industries are increasingly adopting water reuse practices to reduce water consumption and improve sustainability. Advanced treatment technologies, such as membrane filtration and reverse osmosis, enable the recycling of process water and wastewater.

Integrated Water Resources Management (IWRM)

Integrated Water Resources Management (IWRM) is a holistic approach to water management that considers the interconnectedness of water, land, and ecosystems. IWRM promotes sustainable water use, equitable allocation, and environmental protection.

- **Watershed Management**: Watershed management involves the coordinated management of land and water resources within a watershed to maintain ecosystem health and water quality. Practices such as reforestation, soil conservation, and wetland restoration support sustainable watershed management.
- **Urban Water Management**: Urban water management integrates the planning and management of water supply, wastewater, and stormwater systems in cities. Sustainable urban water management practices, such as green infrastructure and

water-sensitive urban design, enhance resilience to climate change and reduce environmental impacts.

Environmental Monitoring and Data Analytics

Innovations in environmental monitoring and data analytics provide valuable insights into environmental conditions and trends, supporting informed decision-making and effective management.

Remote Sensing and Earth Observation

Remote sensing technologies, such as satellites and drones, provide comprehensive and timely data on environmental variables, including land use, vegetation cover, water quality, and air pollution.

- **Satellite Remote Sensing**: Satellites equipped with sensors capture data on various environmental parameters at global and regional scales. Remote sensing data is used for monitoring deforestation, tracking climate change, and assessing natural disasters.
- **Unmanned Aerial Vehicles (UAVs)**: Drones, or UAVs, equipped with cameras and sensors, offer high-resolution data for environmental monitoring. Drones are used for applications such as wildlife surveys, crop monitoring, and habitat mapping.

Internet of Things (IoT) and Sensor Networks

The Internet of Things (IoT) enables the integration of sensors and devices into networks that collect and transmit environmental data in real time.

- **Environmental Sensor Networks**: IoT sensor networks monitor environmental conditions, such as air and water quality, soil moisture, and weather parameters. These networks provide continuous data for early warning systems, pollution control, and resource management.
- **Smart Cities**: Smart cities leverage IoT technology to enhance urban sustainability and resilience. Smart city solutions include real-time traffic monitoring, energy-efficient street lighting, and adaptive water management systems.

Big Data and Artificial Intelligence (AI)

Big data analytics and artificial intelligence (AI) play a crucial role in processing and analyzing large volumes of environmental data, generating actionable insights and supporting decision-making.

- **Predictive Modeling**: AI algorithms and machine learning models analyze historical and real-time data to predict environmental trends and events, such as climate change impacts, pollution levels, and natural disasters. Predictive modeling supports proactive and adaptive management strategies.
- **Data Integration and Visualization**: Big data platforms integrate diverse environmental datasets from multiple sources, providing a comprehensive view of environmental conditions. Data visualization tools, such as interactive maps and dashboards, enhance the accessibility and usability of environmental data.

The Path Forward: Challenges and Opportunities

While significant progress has been made in environmental science and technology, challenges remain in achieving sustainability and addressing global environmental issues.

Technological Challenges

Developing and scaling up innovative technologies to meet sustainability goals requires overcoming technical and economic challenges.

- **Scalability and Cost**: Many sustainable technologies face challenges related to scalability and cost-effectiveness. Continued research and development, along with supportive policies and incentives, are needed to drive technological advancements and reduce costs.
- **Integration and Interoperability**: Integrating diverse technologies and systems to create cohesive and efficient solutions can be complex. Interoperability standards and collaborative approaches are essential for successful implementation.

Policy and Regulatory Challenges

Effective policies and regulations are critical for promoting sustainability and supporting the adoption of innovative technologies.

- **Policy Coherence**: Ensuring coherence and alignment among policies related to energy, water, agriculture, and environmental protection is essential for achieving sustainability goals. Integrated policy frameworks and cross-sectoral coordination are needed.
- **Regulatory Barriers**: Regulatory barriers can hinder the deployment of sustainable technologies and practices. Streamlining regulatory processes and removing barriers can facilitate innovation and adoption.

Social and Behavioral Challenges

Achieving sustainability requires changes in social norms, behaviors, and practices.

- **Public Awareness and Education**: Raising public awareness and understanding of environmental issues and sustainable practices is essential for driving behavior change. Education and outreach programs play a critical role in promoting sustainability.
- **Equity and Inclusion**: Ensuring that the benefits of sustainable technologies and practices are accessible to all communities is crucial for achieving social equity. Inclusive approaches and targeted interventions are needed to address disparities and promote equity.

Opportunities for Collaboration

Collaboration among stakeholders, including governments, businesses, academia, and civil society, is key to advancing sustainability and addressing global environmental challenges.

- **Public-Private Partnerships**: Public-private partnerships can leverage the strengths and resources of both sectors to drive innovation and implement sustainable solutions. Collaborative initiatives can accelerate progress and achieve shared goals.
- **International Cooperation**: Global environmental challenges require international cooperation and coordinated action.

Multilateral agreements, international organizations, and cross-border collaborations are essential for addressing issues such as climate change, biodiversity loss, and pollution.

Innovations in environmental science and technology are paving the way for a sustainable future, offering new tools and approaches to protect our planet and ensure the well-being of future generations. From renewable energy and sustainable agriculture to waste management and water conservation, these advancements are transforming how we interact with the environment and use natural resources.

As we move forward, it is essential to continue investing in research, development, and innovation, while addressing the challenges and barriers to sustainability. By fostering collaboration, promoting equity, and implementing effective policies and practices, we can build a sustainable future that benefits both people and the planet.

The journey toward sustainability is ongoing, and the innovations of today will shape the world of tomorrow. As we embrace new technologies and approaches, we must remain committed to the principles of sustainability and work together to create a healthier, more resilient, and more equitable world for all.

Milton Keynes UK
Ingram Content Group UK Ltd.
UKHW010626190824
447136UK00010B/151